John Thompson Dickson

The Science and Practice of Medicine in Relation to Mind

The pathology of nerve centres and the jurisprudence of insanity, being a course of

lectures delivered in Guy's Hospital

John Thompson Dickson

The Science and Practice of Medicine in Relation to Mind
The pathology of nerve centres and the jurisprudence of insanity, being a course of lectures delivered in Guy's Hospital

ISBN/EAN: 9783337315412

Printed in Europe, USA, Canada, Australia, Japan

Cover: Foto ©berggeist007 / pixelio.de

More available books at www.hansebooks.com

THE SCIENCE AND PRACTICE

OF

MEDICINE IN RELATION TO MIND

THE

PATHOLOGY OF NERVE CENTRES

AND THE

JURISPRUDENCE OF INSANITY

BEING

A COURSE OF LECTURES DELIVERED IN GUY'S HOSPITAL

BY

J. THOMPSON DICKSON, M.A., M.B. (CANTAB.)

LECTURER ON MENTAL DISEASES AT GUY'S HOSPITAL; LATE MEDICAL SUPERINTENDENT
OF ST. LUKE'S HOSPITAL.

"On earth there is nothing great but man.
In man there is nothing great but mind." SIR W. HAMILTON.

NEW YORK:
D. APPLETON AND COMPANY,
549 AND 551 BROADWAY.
1874.

TO

PHILIP CAZENOVE, ESQ.,

A GOVERNOR OF

GUY'S HOSPITAL, AND ALSO OF ST. LUKE'S HOSPITAL,

IN APPRECIATION OF

HIS ADMIRABLE PERSONAL CHARACTER,

AND IN GRATEFUL ACKNOWLEDGMENT OF MUCH KINDNESS,

THIS WORK IS DEDICATED BY

THE AUTHOR.

PREFACE.

These Lectures were first delivered in Guy's Hospital in the summer of 1869. They have of necessity been considerably modified since that time, and are now published as they were delivered in the summer of 1873, and are in a great measure, a record of personal observation.

I have striven, as far as possible, to teach simply what I know of my own knowledge, as I believe that the most successful teaching is that which is based upon personal experience.

In collecting materials in the first instance, I had recourse to every available work and pamphlet on the subject of insanity, and I availed myself of the successful labour of numerous cultivators of mental science. The *Journal of Mental Science*, offered a wide field wherein I found vast masses of information; and the admirable handbooks of Drs. Bucknill and Tuke, of Griesinger, (translated by Dr. Lockhart Robertson and Dr. Rutherford,) the Lectures of Dr. Blandford and the excellent Work of Dr. Maudsley, were all on my desk whilst preparing these lectures. I must here acknowledge my indebtedness to the several authors, and, generally, to any whose name I may have inadvertently omitted to note.

The pathological anatomy, which forms a considerable part of the practical teaching, is almost entirely new. In delivering the lectures it was illustrated by a large collection of microscopical and other preparations and drawings, plates of some of which I have inserted; they were drawn with the utmost care from the microscope, and are faithful representations of the morbid appearances. I would take this opportunity of thanking my friend, Dr. Lockhart Clarke, to whose patient teaching I owe whatever of success I may have achieved in making microscopical preparations, and also in describing and demonstrating nerve-tissue, for the great assistance he has from time to time so liberally afforded me.

I have given the pathology of nerve centres a prominent place in this work, because I felt that the only successful manner in which mind could be studied, was in its relation to organization, and that the surest means of comprehending abnormal conditions of mind, was by gaining a good knowledge of the abnormal tissues upon which it depended.

Pathology is a progressive science, and nervous pathology is yet in its infancy. I think, nevertheless, that my descriptions of morbid structures will bear the test of the most severe criticism. But facts in pathology, as in every other science, must be collected before we can generalise and draw inferences from them, and I therefore ask for my descriptions, nothing more than that they be received as a record of facts and observation.

I have included in the last lecture some suggestions for a natural classification of insanity. It will, I think, be found to be purely scientific. I have endeavoured to render it simple, and at the same time comprehensive.

To my erudite friend, Dr. Diamond of Twickenham, my best thanks are due, not only for many useful hints, but for his liberality in placing at my disposal, his valuable collection of photographic negatives, some of which have been reproduced by a new process of litho-photography. The work was executed by Mr. Briggs of Baker Street.

I have to offer my best thanks to Dr. Stocker of Peckham House, to whom I am indebted for the opportunity he has afforded me of practically instructing my class in his Asylum.

The excellent arrangement of Peckham House, which has lately been reconstructed, afforded the students the best possible opportunities of observing insanity, and the benefit of a well-ordered asylum in the treatment of insane patients.

Finally, I have to thank my friend Dr. Moxon for

his kindness in assisting me in correcting many of these sheets as they passed through the press, and for the many valuable suggestions he made to me whilst the work was in progress.

26 Upper George Street,
Bryanston Square, W.
January, 1874.

CONTENTS.

PAGE

LECTURE I.

Introduction—Mental Disease—Object and Subject of Study—Grounds upon which the Principles of the Study are based—Plan of the Course—Outline of Pathological Anatomy of Insanity—Outline of Pathology, Etiology, Classification and General Morbid Anatomy of the Insane—Complications of General Disease with Insanity—Diagnosis—Popular notions of Insanity and Popular objections to Medical Views of Madness—Outline of Prognosis—Outline of Treatment 1

LECTURE II.

THE RELATION OF LUNATICS AS TO LAW.

Legal responsibility of Lunatics—Proceedings in Chancery—The Lord Chancellor's Visitors—The Lunacy Commission—Certificates of Lunacy—General directions for filling up and signing—Persons prohibited from signing Certificates—Certificate and Order for Pauper Patients—Order and Statement—Regulations for the Reception of Lunatics—Single Cases—The Medical Attendant and his Duties—Statement to be made after two clear days—Medical Visitation Book—Special arrangement for Patients under the care of a Medical Man—Medical Journal—Notice of Death to Coroner and to Commissioners—Notice of discharge, escape and recapture—The order for discharge, transfer and removal—Leave of Absence—Annual Report—Licenses—Lunatic Hospitals—Private and Public Asylums 36

LECTURE III.

ACUTE MANIA.

Insanity generally—Definition of Insanity illustrative of Acute Mania—The term Mania, its definitions considered—Temperature Pulse—Secretory and Excretory Organs—Excitement—Delusion—Cases of Acute Mania—Natural History—Prognosis—Photography in illustration of Mental Diseases—Etiology of Acute Mania—Pathology of Acute Mania—Imperfect Recovery passing on to various states of Chronic Mania 70

LECTURE IV.

CHRONIC MANIA.

Chronic Mania continued—Cases—Monomania So-called—Varieties of Mania—Epilepsy and Mania—Various Phases of Mind Associated with Epilepsy—Pathology of Epilepsy—Sleep-talking and Sleep-walking—Catalepsy—Demonomaniacs 111

LECTURE V.

VARIETIES OF MANIA (*Continued.*)

Varieties of Mania so-called—Primary and Secondary conditions—Symptomatic Mania—Recurrent Mania—Accidents—Insolatio—Cardiac Disease—Puberty—Climacteric Period—Masturbation—Puerperal State—Menstruation—Metastatic Insanity—Phthisis—Syphilis—Acute Disease—Rheumatism and Fever—Gout—Habitual Discharges and Cutaneous Eruptions—Pathology of Chronic Mania—Treatment of Mania 137

LECTURE VI.

MELANCHOLIA.

Melancholia Described—Derivation of the Term—Definition of Melancholia—Forms of Melancholia—Cases of Melancholia—Post-mortem Appearances—General Condition of Melancholia—Sleeplessness—Exaggeration of Impressions—Hallucination in Melancholia—Morbid Religious Impressions—Sudden Impulse—Masking of Feelings Suicidal Tendencies—Bodily Condition—Varieties of Melancholia—Pathology of Melancholia 177

LECTURE VII.

ETIOLOGY OF MELANCHOLIA.

Heart Disease and Melancholia—Post-mortem Condition of Various Viscera—Syphilis—Diagnosis of Melancholia—Treatment of Melancholia—Rational Principles—Surroundings—Food—Feeding—Mechanical Feeding—Stomach Pump—Nasal Tubes—Nose Feeding—Drugs—General Paralysis of the Insane 212

LECTURE VIII.

Progressive Paralysis and Insanity.

Alienist's definition of General Paralysis—Dementia Paralytica—Progressive Paralysis—Exaltation of Ideas—General Paralysis without Delusions—Early Symptoms—Duration and Progress—Varieties of General Paralysis—Muscular Paralysis Indefinite—Epileptiform Attacks—Pulse—Paralysis of Cranial Nerves—Sphincters—Age of General Paralytics—Progressive Paralysis in Women—Pathology—Brain—Cord—Sclerosis of Nervous Tissue—Progressive Locomotor Ataxy—Atrophy of Nerve-Tissue—Sympathetic Ganglia—Causes—Prognosis—Treatment 243

LECTURE IX.

Dementia.

Dementia Defined—Comparison between Dementia, Imbecility and Idiocy—Acute Dementia Illustrated—Pathology of Acute Dementia—Causes of Acute Dementia—Chronic Dementia—Forms—Sequel of Acute Mental Disease—Primary Dementia—Infantile Dementia . 289

LECTURE X.

Idiocy and Imbecility.

Idiocy—Imbecility—Heads of Idiots and Imbeciles—Conformation of Mouth and Palate—Development of Brain and Skull—Imbecility—Liability of Idiots and Imbeciles to Mania and Melancholia—Dumb Paralytics often mistaken for Idiots—Diagnosis and Prognosis from Mouth and Palate—Etiology—Syphilis—Fright to Mother during Pregnancy—Drunkenness in Parents—Marriage of Consanguinity—Ethnical Classification of Idiots—Treatment of Idiocy and Imbecility 327

LECTURE XI.

Etiology of Insanity.

Disease of Mind dependent upon Morbid Tissue—Hypothesis of Possession of the Devil—Moral Influence—Waking Hallucination—Religious Impressions—Potentiality of Insanity—Drunkenness—Precocity—Influence of Marriage among the Tainted—Syphilitic

Diseases of the Brain Starvation—Blood-supply to the Brain Narrowing of Vessels—Aphasia—The Development of Language—Chloroform—Railway Travelling—Excess of Brain-work—Influence of Emotions—Disappointment—Surprise—Conscience—Artificial Condition of Living—Influence and State of Society—General note on the Pathological Anatomy of Insanity—General note on the Treatment of Insanity 356

LECTURE XII.
MORBID STATES OF MIND.

General symptoms of Morbid states of Mind—Delusion Illusion—Hallucination—Delirium and Delusion—Classification of Insanity—Enumeration of Phenomena of Morbid States of Mind from Insanity of Sense—Insanity involving the Emotions—Sympathetic Depression—Moral Sense and Moral Insanity—Conclusion . . 394

DESCRIPTION OF PLATES.

FIGS. 1—6. SPINAL CORD. ATROPHY.

FIG. 1.

Section natural size, from upper dorsal region.
- α. Indicates the shrunken cord.
- β. The dark mass below, shows an enormous thickening of the dura mater. Pachymeningitis. See pages 278-9.

FIG. 2.

Section natural size. Upper part of lumbar region.

FIG. 3.

Is No. 1, magnified by half-inch power and reduced.
- α. Indicates anterior and β posterior columns, which are the most changed, the tubes being collected into bundles and shrunken.
- γ. Indicates the lateral column which is less wasted.
- γ. The tractus-intermedio-laterales: the fibres of which are indistinct.
- δ. Is the posterior root of a nerve very much wasted.
- ε. A small group of cells; the posterior vesicular column.
- ζ. Indicates the spinal foramen filled with granules, and above it and below it at ι, several small perivascular canals, containing pieces of vessels and hæmatozin. See pages 276 and 277, description of specimen marked No. 2.
- νν. Represents a portion of the pia mater.
- λλ. The cut ends and portions of vessels in the same.

Fig. 4.

Section from upper part of the lumbar region, showing cells in three groups in each of the anterior cornea. The cells have not been reduced to the same extent as the remainder of the drawing, in order to render them more obvious. The posterior roots of nerves are much wasted.

The white tissue is much wasted, but more vessels are seen in this than in No. 3, the section from the upper dorsal region. δ. Is a spot in which the tissues were broken down.

Fig. 5.

Some of the cells from the lumbar region magnified by one-sixteenth of an inch; some of them contain granular matter, others are but little altered.

Fig. 6.

Debris from a small cavity in the middle of the lumbar region.
- α. A crystalline body.
- β. Some broken nerve fibres.
- γ. Rounded bodies amyloid in appearance.
- δ. Granules from broken-down cells.
- η. Rounded or globular bodies; some of them looked like fat, others seemed to be gelatinous.

Fig. 7.—Section of Brain.

Section of Brain: Atrophy; from a case of Progressive incomplete Paralysis with Dementia.

The plate shows convoluted vessels contained within the hyaloid or fibrous sheaths of the perivascular spaces, the sheaths have become disconnected from the brain tissue which is shrunken, leaving very wide canals, upon and within the sheath are numerous crystals of hæmatozin. See pages 271 and 272, also page 209.

This drawing is from section marked G.P.C., described in the text.

Fig. 8.

Section of Spinal Cord. Cervical region, from a case of Progressive Locomotor Ataxy.

Showing posterior columns in the extreme degree of Sclerosis, the change having progressed until the columns have become little else than areolar tissue. The cells of the grey matter are for the most part normal. A few blocked vessels, showing blood-cells and pigment, are visible, and coloured yellow. See page 276.

Fig. 9.

Section of Spinal Cord. Cervical region, from a case of Progressive Muscular Atrophy.

Showing enormously dilated vessels, and a remarkable absence of cells in the grey matter.

The white tissue shows dilated tubes and numerous prominent vessels. See page 277.

LECTURES ON INSANITY.

LECTURE I.

Introduction—Mental disease—Object and Subject of Study—Grounds upon which the Principles of the Study are based—Plan of the course—Outline of Pathological Anatomy of Insanity—Outline of Pathology, Etiology, Classification and General Morbid Anatomy of the Insane—Complications of General Disease with Insanity—Diagnosis—Popular notions of Insanity and popular objections to medical views of Madness—Outline of Prognosis—Outline of Treatment.

GENTLEMEN,—We commence to-day a course of Lectures on Mental Disease. A section of general Medicine separated into a distinct branch of medical science, not because in its various forms mental disease differs *per se* from general disease, but because in its social aspect, the aspect in which we have practically to deal with it, mental disease appears very differently to disease generally. The sufferer from mental disease is not a moral agent, his actions are not made amenable to a standard of right and wrong. He is therefore not a responsible or accountable agent and, differing in this particular from the rest of mankind, he becomes a study in himself.

The expression 'mental disease' is not new to science, and it is sufficiently clear to convey its meaning without explanation, whilst the synonyms Psychiatry and Psychological Medicine which of late have unfortunately been imported into our scientific nomenclature, are in themselves almost delusive, pointing as they do to the ψυχή or soul, of which it is not our province to treat, rather than to the subjective mental phenomena, which properly come under our observation.

The question of the human soul is one which does not belong to the subject upon which we are about to enter, and its importation only complicates a branch of study, which, save for the mystification of Spiritualistic Philosophers, might long since have advanced far beyond its present boundaries.

In every scientific enquiry it is essential to obtain, as far as possible, at the outset, a clear notion of the nature and property of the materials we have to deal with, and in no enquiry perhaps is such a knowledge more necessary than in that concerning the nature of mental disease. It, therefore, becomes my duty at once to place before you as clearly as possible the subject and the object of our study, and so to circumscribe them by generalization that we may have a firmer basis than that of mere speculation to build the superstructure of our science upon.

Firstly then, the object of our study is two-fold, and may be considered under the heads of *general* and *personal*.

The general object is of course the same as that which is common to scientific investigation of every kind, viz. the elimination of truth. Another general object is the advance of a science still struggling in its birth, but one which bids fair to lighten the burden of human suffering and to diminish the grievousness of human ills.

The personal object of our study is to place ourselves in a position to relieve the sufferer, and by assisting those in authority, or who have charge of him, to preserve his rights.

It was felt that the relations of Insanity to Medicine had become so important that a course of the study of the latter was incomplete if it excluded the former, and therefore this supplementary course was instituted.

These lectures will be illustrated by a Practical Clinique to be held at Peckham House, by the permission of Dr. Stocker, the proprietor, who has kindly placed this asylum at my disposal for the purpose of teaching and study.

Secondly—The subject of our enquiry is, the Mind in a condition of disease. According to modern views, it is thought hardly possible that we can comprehend the morbid phenomena we call Insanity, until we have some notion of mental phenomena generally. In teaching it is usual to pass in review the subject of the mind in a state of health before attempting the study of mental disease. My own experience, however, leads me to believe that such a method of procedure is unnecessary, and I feel that any lengthened dissertation upon the subject of the mind, would at the present time fill you with alarms and suspicions and lead you to determine to neglect mental disease altogether: for you might naturally reason to yourselves that if the difficult and indefinite limits of Physical and Metaphysical Philosophy are made to appear the only roads and approaches to the subject of mental disease, you would be wise to hesitate before entering upon them, however pregnant the various subjects might be with interest, and however useful to you in practice a knowledge of the disease might be.

It may be argued *à priori* that as with somatic diseases generally—we cannot comprehend the phenomena resulting from the disturbance of any organ unless we have some knowledge of the functions of that organ in a condition of health (*e.g.* we could not understand the phenomena associated with jaundice unless we knew something of the normal functions of the liver); so with mental disease and the various forms of malady we call madness, we need hardly attempt to define or explain them with any precision until we have some knowledge of mental phenomena in health. But it may also be argued that the functions of the liver could never have been perfectly discovered except under the light that was shed upon the subject by various forms of liver disease, and granting this I would argue *à fortiori* that in the case of the brain and mind, although the mind is not a secretion of the brain in the same sense as that bile is a secretion of the liver (as was suggested by Cabanis), yet we never could have come to a knowledge even of the seat of mental phenomena had we not had a light shed upon the subject by diseases affecting the brain. The disturbance of the mind, attendant however upon many abnormal states of brain has enabled us at least to demonstrate the dependence of the mind upon the material brain; and the contrast of the arrested, disturbed, or perverted movements of the mind under abnormal states of the brain with the calm and sequential performance of the mental faculty under healthy conditions of the brain, has afforded us facts for judgment, and premises for inference as to the nature of mental phenomena themselves. If in the first place we consider the various forms of mental disease, you will by familiarizing your minds with them prepare yourselves for more accurate observations of the mind itself, and then, on attaining a knowledge of the mind in health, it will be possible for you to understand more readily the many and various phenomena which are presented to you as madness. If, too, an interest has to be made for you in the subject we have in hand, that object will hardly be attained by a headlong plunge into the depths of mental and metaphysical philosophy, and I propose, therefore, to give you by way of introduction merely a short statement of the grounds upon which the principles of our study are based; we shall then consider, briefly, the outline of the Pathological anatomy, the Pathology, Etiology, and Classification, of mental disease; the general pathological condition of the insane, and the Diagnosis, Prognosis, and Treatment of Insanity. We shall then discuss the relations of Insanity to law, and the manner of filling up the forms of order and certificates. After which we shall pass on at once to the illustration and consideration of the subject of madness, reviewing the physical and less obvious conditions in juxtaposition with the more prominent, or as they have been termed, the mental side of the phenomena. And I shall reserve until the end of our course the enquiry into the subject of the mind itself. I shall then place before you a *resumé* of mental phenomena generally, and at that stage of our study, rather than now, the subject will interest you, and you will find that it will throw a new light upon many of the phases of mental disease which you will by that time have seen and thought about.

The grounds upon which the principles of our study must be based are those of a material philosophy. Whatever be the force of the Socratic arguments, and the power of the Berkleyan philosophy, we shall find ourselves lost in a boundless and unfathomable ocean of speculation if we go beyond the brain for explanations of the phenomena of thought and madness. Whatever may be the shortcomings of material philosophy, and I shall, later on in the course, discuss the subject, one point is certain, viz. that whilst considering mental diseases under the light of material philosophy the study has been advanced, whereas on the other hand, so long as it was regarded through the medium of the many and fanciful systems of philosophy which have been built up and demolished, so long was it fettered and shackled and so long was its progress or advance impossible. In a word, we shall regard all mental phenomena as the result of changes in the cells of the brain or the resultant of forces acting on or bearing upon the brain cells. Whilst we shall regard consciousness as the result of a simple change in the condition of a brain cell, we shall regard self-consciousness as the comparison of two or more changes of impressions which have occurred either in the same cells or in different cells.

If mental health be the calm and sequential performance of the function of cerebration, mental disease is the disturbance or arrest of that calmness or the interruption of that sequence ; and if the former be the ordinary and necessary phenomena which occur in the grey substance of the cerebrum during health, it is but logical to conclude that when the latter, viz. the disturbance, arrest, or interruption occurs, it results because the grey substance of the cerebrum is diseased. Insanity is, however, something which is to be seen rather than described, and it is to be recognised by its clinical features rather than by its fulfilling the strict requirements of definition. By and by we shall consider how healthy mental phenomena are dependent upon healthy conditions of brain, and we shall see how these same phenomena are varied by disease. But what we have now to consider is the expression in subjective phenomena, of objective disease of mind. I would, however, again remind you that the objective diseases of mind are not different *per se* from diseases generally, but that it is the manifestation of these diseases in the modification of the ordinary subjective phenomena of mind which calls for the special and separate consideration of mental disease as a section or class in practical medicine.

Outline of Pathological Anatomy.

If we consider, in the first place, the pathological anatomy of the insane, we shall find, speaking generally, that it does not differ from the pathological anatomy of the sane, though we find structural changes in the brain of those who have lived as insane beings which we believe to be incompatible with the due performance of mental function.

Wasting or atrophy of the brain are, perhaps, the most common conditions which are found on the *post-mortem* table in association with insanity, and necessarily coupled with these are thickening of the membranes, œdema, and aqueous effusions. Sometimes we find deposits of pigment, hæmatin, amyloid cells and calcarious particles, widening of the pervascular canals and thickening of the bones. Sometimes induration of the white substance with flattening of the grey convolutions is found, exhibiting increase of areolar tissue with destruction of the nerve-element, also fatty degeneration, and as an accompaniment of this state of things the skull is sometimes found to be abnormally thin. Sometimes growths of bone project into the skull, sometimes osseous plates are formed in or on the membranes. Sometimes white or œdematous softening of the cerebral mass is the condition found; sometimes yellow softening, and occasionally, though rarely, red ramollisement is found. Often the vessels, especially the arteries, at the base of the brain are found to be atheromatous; sometimes the vessels are injected, and sometimes hæmorrhagic effusion into the substance of the brain is found; sometimes cicatrices of old hæmorrhages, or tumours, or cysts, or adhesions of parts that should be separate are present; sometimes cysticerci are found, sometimes excess of fluid is found in the ventricles especially when the brain is wasted; sometimes pus is found, and sometimes the ependyma, or lining membrane of the ventricles, the septum lucidum, or the arachnoid membrane is covered with crystaline granulations indicating that an inflammatory process has, at a not very distant date, occupied these structures. But any of these conditions may be, and many often are, found in the necrops of persons who have maintained in more or less perfection their mental powers to the last moments of their lives. And we must remember that, notwithstanding the immense value and importance of pathological anatomy, yet pathological anatomy is not pathology, and as in general diseases, so in insanity, the anatomical lesions which are found after death may be, and probably often are, more the result of the pathological processes which during life was expressed in the subject phenomena, than themselves the cause of the symptoms.

Thus there may from some cause be an arrest in the nutrition of the brain, the subject phenomenon of which will be some form of insanity, and the object phenomenon will be wasting or atrophy of the brain. In such a case it would be incorrect to say that the insanity was the result of cerebral atrophy.

On the table are specimens and drawings of many of the various pathological conditions I have enumerated and as we proceed I shall specially illustrate as far as I am able each part of our subject with microscopic and other preparations, photographs, drawings, and specimens.

Although definite structural cause can not always be absolutely defined,—although the microscope has not demonstrated one special alteration in the brain which can always be recognised as the cause of the subjective phenomena,—and although we cannot point to a brain and say with absolute certainty from the conditions presented to our eye, "that is the brain of insanity," in the same manner as we can point to the lungs of a subject of phthisis pulmonalis and say, from characteristic appearances, those are the lungs of phthisis, yet there are not only good grounds for believing as some have asserted, but there is no ground for disbelieving, that in the case of insanity some change either coarse or fine has occurred in the brain; and perhaps with better reason we may affirm that a change has occurred in the brain in a case of insanity, than in the lungs of a patient suffering from tuberculosis, for in the ordinary relation of effect to cause, we cannot have variation in the former without alteration in the latter,—we cannot have variation in the subject mind without some, no matter how inappreciable, change in the object brain. But we may have phthisis latent in the constitution without any local expression in the lungs. The late Professor Schrœder Van der Kolk expressed a strong opinion that a change capable of being demonstrated always occurred in the brain of an insane person and I believe he was correct in his opinion. As a rule however, if you find on the *post-mortem* table a wasted brain, and evidence of either old or recent sub-acute inflammation of the membranes, you may conclude with almost moral certainty that some mental defect existed.

Outline of Pathology.

The Pathology of Insanity consists in a greater or less degree of arrest of the function of that portion of the brain wherein cerebration occurs.

The characteristic of some forms of insanity is excitement and vividness of impression, but this excitement and vividness always emanate from one portion or spot of the brain which pays out its functional activity rapidly and unchecked or uncorrected by the portions whose functions are arrested or in abeyance. In the healthy mind one impression or one set of impressions acts as a check upon another, and so our impressions mutually correct, balance, compensate, and control one another; but if the function of one portion of the brain is arrested or in abeyance, the function of the active portion is performed without check, correction, balance, compensation or control, and according to the degree of its activity it may expend itself rapidly like a clock without a pendulum, or a watch without a hair spring, and from want of correction vivid pictures may be exhibited by active cells, the ideas imprinted upon the brain may assume any degree of grotesqueness, and the excitement may assume any degree of intensity. According to the degree of arrest too, so will be the diminution of mental function, and it may range from a slight impairment of the power of reasoning upon one subject down to absolute dementia. When the brain is quite healthy its functional activity is perfect and the capacity and power of the mind is proportionate to the quantity of the brain matured. When the brain becomes unhealthy the functional activity becomes imperfect, and the capacity and power of the mind will fail in proportion to the degree of unhealthiness and change and arrest of function which occurs.

Outline of Etiology.

There cannot be a question as to the importance of the study of the cause of insanity: the place which it fills is perhaps second only to the manifestation of the disease itself.

The importance of the subject may be estimated when I tell you that the number of lunatics who come under official cognizance increases at the rate of about 2000 a year, and that on the 1st of January, 1872, there were no less than 58,810 lunatics on the books of the Commissioners, exclusive of numbers who are under the special supervision of the Lord Chancellor's visitors, and those at home with their relations.

As in general diseases, so in diseases of the mind, the etiology divides itself into

A. Predisposing, and B. exciting causes—and these we must briefly consider.

The actual cause of insanity, in all cases, is undoubtedly malnutrition of the brain, but etiology seeks to know something more and looks for a cause for this malnutrition.

The operation of predisposing causes is markedly seen in insanity, and perhaps the most powerfully marked is hereditary tendency. Exciting causes also, both physical and moral, are often very well defined and illustrated.

The question has often been asked whether or not a potentiality* of insanity is necessary, and from my observation I hold with the opinion that it is necessary, but it should be clearly understood that such a potentiality may be developed in anyone, as for instance a blow on the head may so disorganise the brain that the person in consequence may become insane.

A. Under the head of predisposing causes we must enumerate

Hereditary taint and epilepsy.

Marriages of consanguinity.

Accidents, as blows on the head, shock, wounds

* A possibility peculiar to the individual; or a quality which exists *in potentia*.

and cuts, cerebral hæmorrhage, sunstroke, tumours in the head, syphilis, heart disease, fever, together with the various forms of retrograde metamorphosis of the brain already noticed, rheumatism and chorea, also starvation, and it has been said, in some cases, continence, particularly in women.

B. Under the head of exciting causes we have two classes, Physical and Moral.

Amongst the physical exciting causes are exhaustion from any cause, as epilepsy, the immoderate use of alcohol, pregnancy and childbed, railway travelling, and exhaustive disease as tuberculosis and cancer and uterine disturbance, to which may be added onanism and excess of venery.

The moral exciting causes are excess of brain work or any undue tax upon the brain through the medium of the mind. These may take their origin in excess of study, disappointment, fear, love, the pain of stricken conscience, or uncertain ethics and religious excitement, to which may be added the intermediate effects of want and privation, and want of rest.

Any of these causes may bring about a condition of insanity, and in the present day whilst we are living at railroad speed and whilst everyone is making haste to get rich, whilst competition runs high, whilst want and privation are the necessary lot of thousands, and the chances of fortune or ruin daily tremble in the balance with thousands more of the trading and speculating portion of the community, it is not to be wondered at that some of these should wander and depart from a normal and healthy mode of mind and thought. But sometimes the origin of mental disease is very obscure and its etiology very difficult. The question of cause has been compared with that of the obscure cause of the drooping of a percentage in the community of leaves on a tree before their time; but the comparison is inapplicable, though the simile in the comparison of insanity with fading and drooping leaves is remarkably apt. A faded or half-dead leaf is a useless, painful, sad and depressing object, so is an Insane person and he may be considered almost as useless to society, as a fading leaf is to a tree.

Hereditary taint is a predisposing cause you must be ever on the watch for, the occurrence of it is very common though it is often difficult to discover, but its discovery will aid your diagnosis, help your prognosis, and assist you in your deliberation as to treatment.

Outline of Classification.

The common division of the forms of mental disease is based upon the symptoms and the more prominent evidences, and is contained under six heads, and these as enumerated in the *Nomenclature of Disease* published by the College of Physicians I shall, with slight modification, adopt in these lectures as a natural classification. The diseases are grouped as follows:

MANIA, MELANCHOLIA, GENERAL PARALYSIS, DEMENTIA, IMBECILITY, IDIOCY.

This natural classification, however, is incomplete, inasmuch as it cognates only the disorders of the intellect; I shall therefore place before you a classification of Insanity more extended in its scope and based upon the modern view of the constitution of the mind.

Outline of General Pathological Anatomy.

The general Morbid Anatomy and Pathology of the Insane, the complications and the effects of Morbid processes on their various organs differ little, if at all, from those which occur in the sane. The insane are liable to the same diseases as the sane, and they have not, as some have supposed, any immunity from physical or somatic disease by reason of their insanity, in fact from imperfect innervation they are rather more susceptible of diseases than others, especially those diseases dependent upon an ebbing condition of nervous vitality or *vis vitæ*.

It is very common among insane patients to find boils, carbuncles, thecal abscesses, and sloughs; the skin in particular is peculiarly liable to slough, especially the skin of the extremity as the hands and feet. I have often seen the epidermis rise as though blistered and the dermis beneath it pour out a thin ichorous or purulent fluid highly offensive and so persistent that the part has only healed with coaxing and great difficulty. The difficulty of treating such sores you will find will be increased by the irritability of the patient who as a rule is restless and will not tolerate any surgical interference. In fact you will find surgical treatment among lunatics, generally, a matter of very great difficulty as the patients will not keep on bandages, splints and other appliances. A maniac with a broken leg for instance is always the subject of grave anxiety in an asylum, because you cannot induce him to keep his limb quiet, and it will often be months before the broken ends of the bones of such patients unite, and often the repair when accomplished will be very ugly and look like very sorry surgery.

The most ordinary and common lesions found in the thorax of insane persons are congestion, hepatization and gangrene of the lungs, also tuberculosis, and not uncommonly cirrhosis. The heart too, you will find variously affected, an association also between the state of the heart and the insanity, not infrequently having existed. Opinions vary greatly as to the frequency of heart disease, but I believe it to be much more frequent than recent writers on the subject allow.

In the abdominal cavity you may find lesions of any structure, a striking condition sometimes found is that of dilatation of the stomach or colon, intestinal ulceration too is not uncommon in the insane.

The hygienic condition of asylums generally is good, and the patients freed from anxieties and cares, and placed, as they are, under the most favourable conditions for living long lives, often attain a fair age; but it has not been shown as it has lately been affirmed that the insane live longer than, or even as long as the sane, whilst there are some records to show that many who have been celebrated for the power of their intellects, and have retained their high class mental powers intact during the whole of their career, have enjoyed more than the average duration of life. Some statistics were published some time ago in the *Times* showing that a number of great lawyers who had been noted as men of rare intellectual power, had enjoyed exceptionally long lives: Lord Brougham, as an instance, died at the age of 84.

Of the complications of general disease with insanity, the most important are epilepsy and paralysis, sometimes chorea, and sometimes rheumatism, but these we shall touch upon again, and speak of more fully in their proper place.

Diagnosis.

The diagnosis of insanity in its common forms is not generally one of difficulty, although you will sometimes be led to question whether you are dealing with a case of insanity or a case of general disease. You will also be called upon to diagnose insanity and criminality, and between real and feigned insanity, and the grounds of your diagnosis will be severely criticised by the friends of patients, and more severely by barristers in law courts, and by the public press. The cases of real difficulty are the obscure forms of insanity, out of which questions in law courts so often arise, and which lawyers demand shall be distinguished by clear and definite outlines from criminality, or from eccentricity, as the case may be. The demand of the lawyers and the public is, however, sometimes unreasonable. Insanity, like fever, may often be recognized by the eye of the physician, whilst no demonstration in words would make the fact patent to the unskilled. The overt act of insanity is the only evidence of insanity which is plain to the legal mind, but to the physician it is only one of many evidences which may or may not occur. Insanity is, as I before remarked, something which is to be seen and recognized rather than described. Its manifestation is a dynamical condition which like the lightning's flash may appear, work a work of havoc, and be seen no more.

Fever and Delirium Tremens are the most common forms of general disease, that present mental phenomena allied to insanity.

The actual states of the brain are probably very much the same in Delirium Tremens and fever as in insanity: the difference which exists being in the etiology and in the duration.

Cases of malingering or feigned insanity sometimes present themselves most unexpectedly, and have to be diagnosed, and they rank amongst the most difficult in which the skill of the physician can be exercised.

Eccentricity and Hysteria also have to be considered in the diagnosis of insanity.

It will be useful at this point to illustrate some general diagnosis. You may be called to see a patient said to be raging, destructive and extravagant, and you go into his room you find him excited and walking about, he comes up to you stares at you perhaps addresses you by name, or perhaps mutters some incoherent sounds—he probably is not easily persuaded, and refuses to go to bed, or he will pour out a volley of abuse at one time coherent, at another incoherent, and you might think him mad, were it not that a pungent burning skin, a very rapid pulse, irregular and laboured respiration, and a furred tongue, will lead you to hesitate before you express too decidedly an opinion. The patient's temperature may help you, but you will hardly be able to take it. One evening some time since, I received a message from the wife of a medical man with whom I was well acquainted asking me to see her husband without delay. I went to his house at once, where I found him walking about in a state of great excitement, and my entrance was the occasion for an outburst of personal abuse; after some time, during a short interval of calmness, I succeeded in inducing the patient to get into bed, when he immediately recommenced abusing me, spitting in my face and endeavouring to strike me; after a time, he became calmer, and I was able to examine him, and found his skin pungently hot, the bases of both lungs solid, his pulse very rapid, his respirations laboured, and his tongue dry and red, and I learnt from his wife that he had complained of shivering and malaise for two or three days previously. The case then appeared to me to be clearly febrile, and after the subsidence of the maniacal delirium which persisted with greater or less severity until the following morning, the patient passed through a severe attack of pneumonia, associated with a typhoid form of fever of a very low class. I do not think that you will be very likely to mistake such a case, but you will certainly see fever, especially typhus fever, with maniacal symptoms, and it is of great importance to estimate the meaning of such symptoms when they appear, for you will find that sometimes the fever will disappear and the maniacal symptoms persist. Sometimes the insane patient will pass from a condition of maniacal excitement to one resembling typhus delirium, and the patient will die from exhaustion, whereas if you see such a case for the first time you may certainly be led to the belief that you are dealing with a malignant fever.

The following case will present a not inapt comparison. I was lately asked to see a patient engaged in business as a silk mercer in the west end of London, and on calling at his house I was shown into a room where he was seated in front of a fire. He rose as I entered and stared at me, when I said to him that I had been asked to see him as he was not well. He shook hands and then turned upon his wife, who was in the room, and upbraided her for not having prepared him for my visit. It was necessary to improvise means for calming him, and I assured him that his wife had not any intimation as to the time when I should call, an excuse he was at once quite satisfied with; in fact he seemed very friendly and quite ready to be satisfied with whatever I said. On interrogating him he said that he was comfortable and quite well, though his wife thought him ill, and he believed that the wife, who was suffering from bronchitis, had much more need of a doctor than he. His pulse was very rapid, his skin was hot and his tongue was dry and red; he was shivering and said he felt cold. He persisted that he was quite comfortable and well, and chatted pleasantly on various subjects, but all the while he was trembling, and he evaded the questions I pressed, asking him how could he be comfortable whilst he was shivering from cold? He then laughed and said that he had had something, and I asked him what he had had, but only received in answer the expression, "That's the point." He then turned and stared full at me and asked "what?" After a little while I elicited in answer to questions (perhaps they partook in their character somewhat of the nature of those which lawyers call leading) that "somebody in the house had put poison into the water." On asking him how he knew that the water was poisoned he became very much excited, and said that he had a process of analysis by which he tested it and he volunteered to show me his process and hastily left the room; after ten minutes absence he returned and stated that his hand was too unsteady to allow him to perform the experiment. He had very marked and unmistakable febrile symptoms, but he had not the malaise of fever neither had his case the history of fever. I was unable to ascertain his temperature, but his symptoms had lasted a month without having exhausted him, and I pronounced an opinion that the man was suffering from acute mania. I afterwards learned from several inmates of the house that he was, at times, so violent that they were afraid lest he should cut his wife's throat and then his own, which he threatened to do.

Febrile symptoms often accompany mania, and their intensity is often an index of the acuteness of the mental disease; the temperature in such cases is sometimes higher than fever, sometimes it is as high as 105, though as a rule the increase of temperature in insanity is not more than from 1° to 2° F., above the patient's normal standard. You will sometimes see very acute mania of very sudden invasion, with hot skin, rapid pulse, coated tongue, sordes and aphthæ on gums and lips, profound sleeplessness, noisiness, and violence; of course in such a case where there is doubt you must watch and wait. Carefully guard the patient from harm and a few hours will in all probability throw some light on the condition.

The condition known as delirium tremens is generally, but not always easily distinguishable from maniacal delirium, though the character of the delirium, in the former one in which the patient is easily persuaded, rendering it as it were a class, is the only distinction to be drawn between its mental symptoms and those of insanity.

In delirium tremens the patient frequently has hallucinations of sight; he sees repulsive animals in his room or on his bed, and is affected with visions of horrors, but he is as a rule tractable and easily persuaded, and a little watchful care with the history of the case will help you to a right judgment of the case, before you consign the patient to an asylum. In all cases the history and the duration of the attack must guide us in our diagnosis, and whether the persons be raving in delirium of fever, or suffering from delirium tremens, or the victims of mania, they are equally for the time being of unsound mind, and require care, surveillance, and supervision.

It is very necessary however, clearly to diagnose between delirium tremens and insanity, otherwise you may sign a certificate and incarcerate a person who could be as well, or sometimes would be better treated at home, and upon whom you might unnecessarily brand the social slur of having been in an asylum or having been insane. The history, the fever, the tongue, the temperature, the pulse, the skin, and the duration of the case will help you, but you must also be careful to distinguish between the man who drinks because he is mad, and the man who is mad because he has imbibed too freely, or too long. The latter will if he can, often tell you the truth as to his drinking habit, the former, probably, will try to conceal it. There is one fact connected with the association of insanity and acute disease, which I should here mention, when the latter supervenes in the subject of the former, it is very common for the insane symptoms to disappear entirely. I have seen a maniacal patient become affected with acute disease, as rheumatism or pneumonia or acute phthisis, during the course of which the mania has subsided and exciting delusions have disappeared, and I have observed the reappearance of the mental symptoms on the subsidence of the somatic disease; in such cases the history alone can guide you, and in your practice generally you will find that history is the greatest aid you can have in forming a diagnosis.

The value of history in the diagnosis of feigned insanity is peculiarly great. Feigners of insanity usually assume the dissemblance for a definite end, but commonly break down from an ignorance of the natural history of the particular form of insanity they try to feign. It is by no means uncommon to find dissemblance of insanity among prisoners, and often it is extremely difficult in such cases to form an accurate diagnosis, and the physician even with a perfect knowledge of the natural history of the particular form of insanity dissembled, may be baffled. The conflict of opinion in the celebrated case of Lady Mordaunt is an instance of the difficulty of distinguishing feigned insanity.

Lady Mordaunt was reported to have presented all the symptoms of dementia, even to the entire disregard of personal comfort in attending to the calls of nature, and yet it was maintained that the case was one of malingering.

The patient was afterwards placed under care and treatment as a person of unsound mind, and some time after that, was discharged from her certificates as "not insane," the fact, however, that, shortly after her discharge it was found necessary to place her again under certificates, is a strong corroboration of the soundness of the first opinion.

On the other side of the picture I have for comparison the confession of a hardened vagabond, who was four or five times in one of our County Asylums. The man was by trade a journeyman painter, and in the summer when good wages could be earned with ease he would work, but in the winter when his handicraft was not so much in demand, he preferred the ease and comfort of an asylum to possible competition for labour, and he stated that he found it easy to feign madness. His usual procedure was to stagger against a policeman, and when he had aroused the spleen of the latter by his seeming awkwardness, he would commence to speak of his mansion in the country, he thus got himself taken charge of as a wandering lunatic, and was sent to the workhouse where his intentional violence, and his assumed delusion of his mansion were accepted as evidence of insanity, and secured for him comfortable winter quarters at the expense of the parish.

When a sufficient motive for dissembling is apparent it is necessary to regard the case with the eye of severe criticism, and the more perfect our knowledge of the form of insanity acted, the better will be our chance of discovering the imposture. We should remember that the feigner of insanity often does too much, and so betrays himself. Whilst if a person can successfully play the madman for years, his habits of life render him so akin to the aliens, that perhaps he is in his right place and will do least mischief if left amongst them.

One of the most difficult forms of insanity to recognize and one in which we may often be mistaken is that associated with epilepsy. In this form of disease the alternate recurrence and disappearance of the insane symptoms may be mistaken for attempted deception, and serious results may follow.

Epilepsy is sometimes very transient, but in its most fleeting forms it is often attended with, or followed by the gravest manifestations of mental aberration. In fact the more fleeting the epilepsy, the graver, as a rule, appear, the mental symptoms.

I once saw a case of cut throat in this Hospital, in which the attempted suicide had committed the act during a stage of temporary insanity following an attack of *Le Petit Mal*, but up to the time of the commission of the act, the epileptic nature of the malady from which the girl had long suffered had never been recognised.

The diagnosis and clear demonstration of epilepsy in a case, the subject of a trial, is of the utmost importance as the unequivocal proof of epileptic insanity in the accused ought to be sufficient to annul his responsibility.

In acute dementia, when the patient suddenly ceases to speak, because he has ceased to think, you will be strongly tempted to regard the case as one of feigning, or you may confound it with aphasia; but here again the history will come to your aid, added to which, a little time and careful observation will probably either betray the dissembler or demonstrate the aphasia.

One of the difficulties which you will find in diagnosis is contained in the distinction between eccentricity and insanity. The line of demarcation is often by no means distinct, and perhaps the distinction is in a greater or less degree metaphysical.

Eccentricity is not the result of disease, and like ugliness in the physical features is to be looked upon as a quality rather than a defect. The oddness, by which I mean the departure from conventional deportment in the case of the eccentric, is constant, or habitual and logical; in the case of the insane it is extraordinary and illogical. The famous Dr. Johnson is a commonly quoted and familiar instance of an eccentric man; it is reported of him *inter alia* that invariably on returning to his home, he would strike a number of posts in front of his house with a stick, and if he by any chance missed one, he would retrace his steps, and recommence striking each post in succession, until he reached his door. Now this act, though remarkably odd, was certainly not an insane act, it was habitual and regular, was performed intentionally, and no doubt for a reason which, had the actor explained it, would have been found to be perfectly logical. Possibly it satisfied some arithmetical question in Dr. Johnson's mind; but neither it, nor any other of the Lexicographer's eccentric acts, warped in any degree the powerful judgment of the eminent Author and Satirist; whilst one of the most important evidences of insanity, is a loss of the power of judgment, or the loss of the faculty of correction of an idea, or sensation by comparison with others, in the individuals consciousness or memory.

An insane person who was for some time under my care, had a curious habit of slightly touching with the tips of his fingers everything which he passed; but this would not in itself have constituted unsoundness of mind had he not presented other symptoms. He was however demented, and never spoke, the mental function of his brain having been almost entirely suspended, even to the loss of articulate language, and he lived an automatic existence, eating, drinking, and sleeping, or moving, and touching everything without interest in, or affection for any outward or surrounding circumstances. Here was eccentricity, but loss of reason was evidently along with it.

Hysteria often simulates insanity, and you must not overlook the fact, for you will occasionally be asked for an opinion in cases called hysteria, which you will be unable to distinguish from insanity, and which, notwithstanding the unwillingness of relatives to admit the fact, are cases of insanity veiled under a convenient name.

POPULAR NOTIONS OF INSANITY, AND POPULAR OBJECTION TO MEDICAL OPINIONS OF MADNESS.

You will often find the utmost difficulty in convincing others of the correctness of your diagnosis, and in answering the queries of friends who are unwilling to admit the facts of insanity; and who will urge a series of objections, many of which are the outcome of popular notions, and popular errors. The difficulty too is increased from the fact that the evidences of insanity cannot always be placed in graspable language before the unskilled enquirer. I mentioned just now, that an overt act is the only plain evidence of insanity understood by the public. The insanity of a man "dressing in regal robes, and declaring himself to be a king" is patent, and recommends itself to the minds of lawyers and jurors, or friends and casual observers. They say, here is an act opposed to common sense, and a self-evident delusion, and without further demonstration, they receive it as a fact of insanity; but the case is very different when nothing is patent to the common sense of the ordinary observer, and when it becomes the insanity of the act which has to be demonstrated. I may illustrate this with a case in point. A lady convalescent was under my care; she had been a patient in an asylum, and allowed leave of absence on trial under consent of the Commissioners. This lady had suffered from an acute attack of melancholy; had threatened to commit suicide; and stated as her inducement that she was so ugly, that nobody would fall in love with her. At the time she was sent to me she was convalescing, but was in the feeble condition of mind which is generally to be observed, after an acute attack of insanity has passed away: one day one of her relatives, a lawyer as it happened, called upon me, and was unwilling to allow that the patient had ever been insane at all. He began to discuss her case, and would not be satisfied as to the facts of the patient's insanity, though the lady was too much enfeebled in mind to speak to him when he saw her. He stated the case from his own point of view. He said that he had observed in her exhibitions of temper, such as to cause the utmost distress to aged and invalid parents, whose physical condition called for consideration and gentleness—though at times the patient was kind and gentle to her parents— that at one time she would be loving and amiable with her brother and sisters and friends, but at another time she would not speak to them and only speak bitterly of them; that she would retire to her bed room when anyone she entertained a dislike for entered the house, and that she would shut herself up for days in this bed room, and refuse food; and that when her relatives or friends wrote to her she would re-address and return their letters without opening them; and he desired me to tell him in what this differed from the temper of a naughty child who had always been allowed its own way; and thereby permitted to nurse and nurture selfish and violent passion. Such questions are specious, and it is useless to urge in answer to such queries, that it is contrary to experience to see such a condition of temper grow and develope from childhood, except in cases when the brain is diseased or undeveloped. Conduct and acts such as those described are more than illogical and unreasonable; instances of such dispositions when seen, are usually seen in asylums, and when they occur in an adult of intellectual attainment and good education, and appear as a new phase, they may be taken as *à priori* evidence of an unsound mind.

In the patient who formed the subject of this discourse all the so-called temper was of recent development, and the perversion of feeling and disaffection was conclusive evidence of instability of mind. But sometimes the friends of patients start with a determination not to be satisfied, and will advance puerile objections, such as that insanity is incompatible with the fact that the patients maintain good intellectual power, or that they understand all you say to them. In some cases the intellectual powers maintained by the patients are good, even beyond the average, and the most excited, incoherent, and apparently pre-occupied maniac, will generally understand all you say in his presence, and what is more, remember and perhaps act upon it.

The lady whose case I have mentioned, had given evidence of great intellectual power before her alienation—powers even beyond the average, but the fact did not preclude the possibility of insanity: the mightiest may fall, and the most intellectual may become insane. The emotional part of mind in some patients, may be the part which exhibits disease more prominently than any other; but emotional insanity receives scant justice at the bar of public opinion, until it has succeeded in obviously dethroning reason. You will be expected to explain the difference between emotional insanity and hysteria. Alas! we may almost be tempted to regret that we have such a word as hysteria, so often has it been used as a cloak for ignorance. What is called hysteria is too often insanity expressing itself in emotional disturbance; and if hysterical patients act insanely, endanger their own lives, or the lives of others, or otherwise show themselves beyond the power of their own control, they must be considered as insane and treated accordingly.

If a patient exhibit perversion of the moral sense, especially a defiance of the law of self-conservation, he or she must be regarded as of unsound mind, notwithstanding the name by which her relations and friends choose to call her disease.

You will sometimes have the greatest difficulties in demonstrating insanity, and abnormal mental conditions to juries, and in law courts; the task however when it is to be done, must be done resolutely: when called upon for your opinion, you should arrange your points perfectly clearly in your own mind, before you commence to speak, and you must take care whatever theories you may yourself entertain, that you do not propound any views that may not be comprehended by the common sense of those who have to hear you. A reprimand from a judge is degrading, but to be warned that you are carrying the court into the clouds, is discreditable to a man of intellect and good sense, nor does it tend to raise a Physician in the estimation of the public.

The opposing Counsel will endeavour to put you into a cleft stick, and he will endeavour to induce you to commit yourself to expressions which would illogically draw the general from the particular.

What is insanity in one person is not necessarily insanity in another. Irascibility, or swearing, or obscene language, may be evidence of insanity in a particular person, but neither irascibility, swearing, nor obscene language are general evidences of insanity. You must be on the watch, for in a court of law you may be asked is this, or that, or the other fact, evidence of insanity. It is the Barrister's business to upset your statements, and if he cannot shake the logic of your facts, he will undoubtedly do his best, to lead you into making illogical arrangements of them, and so shake the conclusions which may be drawn from the most obvious and convincing demonstrations.

Outline of Prognosis.

The Prognosis of Insanity is not hopeless, but recovery is often dependent upon early treatment. Patients are often kept at home, or in workhouses until their insanity has become chronic and their recovery almost, if not altogether impossible; whilst, had they been placed under favourable conditions sufficiently early, their recovery might have been certain. The statistics of St. Luke's Hospital, where the majority of cases admitted are, or were recent cases, show that the percentage of recovery in that institution has often been not far short of 70. Such results however, can only be expected in Institutions like St. Luke's, or Bethlehem Hospitals, where the cases are in a certain degree picked cases, and where except under special circumstances, patients are not admitted if they have been insane over a year. A considerable number of cases however, will always recover, and if you have to consider a first attack you may generally look hopefully upon it, unless it has had its origin in some such definite cause, as a blow on the head, epilepsy, or sunstroke. The cases with a simple history of exhaustion are the most favourable, but the degree or the intensity of the insanity, gives you no premiss whatever to ground your opinion upon.

Hallucination of sound should be looked upon as a grave symptom, but it is not hopeless, and though frequently long persistent, it may pass away.

There are on record, cases of insanity which have recovered after twenty years of seclusion; and considering this fact, we rarely need speak hopelessly of any cases we have to express an opinion upon.

Outline of Treatment.

The Treatment of Insanity we shall enter upon fully, as each form of mental disease comes under our consideration, the first question however for the practitioner to decide is, whether or not, the treatment of an asylum is necessary? Sometimes asylum treatment is almost imperative, but in many cases it is almost pernicious. The full consideration of these points we shall resume later on in our course, but the fact that a lunatic requires care and treatment, brings him at once under certain legal restrictions; and as it is illegal for any person to receive, and derive profit from the charge of a lunatic, or an alleged lunatic, without an order and two medical certificates, it will be advisable for me at once to enter at length upon the subject of the relations of lunacy to law in the matter of the order, certificates and other statutory requirements, and the responsibilities and duties of medical practitioners and others in regard to insane patients. These questions shall occupy our next lecture.

LECTURE II.

THE RELATION OF LUNATICS AS TO LAW.

Legal responsibility of Lunatics—Proceedings in Chancery—The Lord Chancellor's Visitors—The Lunacy Commission—Certificates of Lunacy—General directions for filling up and signing—Persons prohibited from signing certificates—Certificate and order for Pauper Patients—Order and Statement—Regulations for the reception of Lunatics—Single cases—The Medical Attendant and his duties—Statement to be made after two clear days—Medical Visitation Book—Special arrangement for Patients under the care of a Medical Man—Medical Journal—Notice of Death to Coroner and to Commissioners—Notice of discharge, escape and recapture—The order for discharge, transfer and removal—Leave of absence—Annual report—Licenses—Lunatic Hospitals—Private and Public Asylums.

In general terms the legal responsibilities of an insane person are extinguished. If it can be shown for instance, that a person executed a deed whilst suffering delirium from any cause, such a deed would be held to be void; and a person committing a criminal act is not held responsible for the act if shown to have been, at the time of its commission, in such a condition of mind that he was unable to distinguish right from wrong. The question of the individual's responsibility of course becomes a question for a jury, and the conflicts that arise in court between doctors and lawyers, on the subject of insanity, are too often exceedingly painful, and a slur upon common sense and the enlightenment of the age.

There seems to be an extraordinary contrast in the manner in which the law regards civil and criminal responsibility in insanity, a difference perhaps to be explained only upon the ground that the one case is tried in a court of equity, and the other in a court of law, and the difference of the light in which these two courts often look at the same question is not only remarkable but proverbial.

The civil responsibility of a person may be annulled and his property guarded with jealous care upon proof of unsoundness of mind, though that unsoundness be only demonstrated by a foolish monomania; but should such a person commit a criminal act, his responsibility is not annulled unless it be demonstrated to proof that the act arose out of his insanity, and that at the time of its commission he was unable to distinguish right from wrong.

It has long been felt by scientific men, and in particular by observers of mental conditions, that the hard and fast line laid down in the legal definition is grievous; and of late some lawyers have shared with medical jurists the opinion that the legal definition of insanity is not in accordance with the knowledge and science of the present age.

The legal view of insanity is based upon what is called a *legal dictum*, and this dictum was propounded in 1843 by Chief Justice Hale in answer to queries put by the House of Lords, to the whole of the judges on the celebrated case of McNaughten, who was tried for the murder of Mr. C. Drummond and acquitted on the ground of insanity.

This legal dictum, known as Hales' Dictum, was

expressed by the Chief Justice in the following terms: "The jury ought in all cases to be told that every man should be considered of sane mind until the contrary was proved in evidence; that, before a plea of insanity should be allowed, undoubted evidence ought to be adduced that the accused was of diseased mind, and that at the time he committed the act he was not conscious of right or wrong."

The legal test of insanity is resolved into the question whether the accused at the time he committed the criminal act knew what he was doing, and whether he knew that what he was doing was wrong; and in this form the rule is usually laid down by the Judges in criminal trials.

Chief Justice Hale insisted that *partial* insanity is no excuse for crime, and he remarked that partial insanity was the condition of very many, "especially melancholy persons, who for the most part discover their defect in excessive fears and griefs, and yet are not wholly destitute of reason; and this partial insanity seems not to excuse them in committing any offence for it is matter capital. It is very difficult to determine the invisible line that divides perfect and partial insanity; but it must rest upon circumstances to be duly weighed and considered by both judge and jury."

Lord Lyndhurst, then Lord Chancellor, followed up in the House of Lords the statement of opinion as expressed by Chief Justice Hale; and he declared that there was no doubt as to the law, that it was clear, distinct and definite, that a man was not to be held judicially insane when tried criminally, if he knew right from wrong. There is a palpable deficiency and want of knowledge of mental disease exhibited in these statements. Many lunatics are able to distinguish right from wrong, whilst they cannot choose between them. There is only one class of case, excluding a few idiotic and demented patients, in which the person is wholly unconscious of his acts at the time of their commission, and that case is the dream-like phase and disturbance of mind attending epilepsy. The demented and idiotic even can rarely be considered as wholly destitute of reason, but the idiot may not know right from wrong.

An epileptic, however, may commit a murder under the influence of the attack, and be absolutely ignorant of his act upon the recovery of his mental balance. But a maniac may commit a murder and gloat over it, and relate the steps of his proceedings with fiendish delight, and all the while admit the wrongfulness of the act. He is, however, not less insane than the epileptic, who commits his murder under the epileptic influence; perhaps the raving maniac is the madder of the two, but there is no standard by which you can compare or measure the degrees. This is the point at which *Hale's dictum* most certainly breaks down. To divide perfect from partial insanity is impossible. A person whose mind is off its balance is insane, the condition is a positive condition. There is no comparison about it; there is none of the element of partial or imperfect in the matter. A man when his mind is diseased is insane, and though the intensity of some of the attendant phenomena may vary, as do the pulse and temperature, unlike these they cannot be measured by numbers or degrees.

Mr. Justice Mellor observed at Leeds (Autumn Assizes, 1865) that he thought the definition of insanity which would excuse from criminal responsibility as given in McNaughten's case hardly went far enough. He agreed with the universal conviction of medical observers that a man might know that he was doing an act which was wrong, and still be labouring under such disease of mind as not to be able to restrain his impulse to do that act; and he allowed that such a man should therefore not be amenable to the criminal law. The opinion of one judge, however, cannot alter the law, and, as lately stated by Mr. Justice Byles, if the law be altered it must be done by act of Parliament.

It is satisfactory, however, to observe that the difficulty appears to be more and more felt, and that a hope of a better legal definition, and one consistent with the abstract idea of law, may be entertained. The abstract idea of law being "the perfection of reason and common sense," the legal views of insanity must be materially changed before they can come up to the test of such a standard.

I may mention, however, that in a trial for murder at Lewes in July, 1872, Baron Martin summed up the case without any allusion to the criterion of responsibility as resting upon a knowledge of right and wrong, and he concluded by remarking that when impulses, which according to the medical evidence they were unable to resist, came upon men, it would be safe to acquit on the ground of insanity.

But, as the responsibility of the lunatic becomes extinguished with his incapacity for obeying the law, so also his right to civil and personal freedom becomes extinguished with his release from responsibility, and as a consequence it is lawful to deprive a lunatic of the control of property, and also of personal liberty. In order however that the lawful protection of a lunatic and his property shall not be made a cloak for abuse, the law has imposed certain formalities and restrictions in regard to lunatics, the neglect of which is punishable as a misdemeanour.

The Lord Chancellor is the legal custodian of the lunatic, and, for the protection and administration of a lunatic's property, procedure in chancery is the only legal mode of relief, and an action will very properly lie against any person who administers the estate of a lunatic without the sanction of the Court of Chancery, though modification in the costly procedure is strongly called for. The procedure in Chancery in the case of a lunatic is called an Inquisition, or a Commission *De Lunatico Inquirendo:* it is an open court held by a master in lunacy sitting as judge, and the cause may, if the patient demand it, be heard by a jury. Evidence of the person's insanity has to be produced; the petition may be opposed by the patient, and he may bring evidence to rebut that which is urged against him. In the event of a decree affirming the insanity of the alleged lunatic, two committees are appointed;

the one to take charge of the person of the lunatic, the other to take charge of his estate, and he is placed under the supervision and inspection generally of the Lord Chancellor's visitors, a board of Commissioners appointed by the Lord Chancellor to pay periodical visits to chancery patients, to report upon their cases, and otherwise to guard their interests. In case a patient who has been pronounced insane after a Chancery inquiry should recover, he can be reinstated in his former position in society and in the management of his own affairs by a process called *Supersedeas*.

But the machinery of a Chancery enquiry is expensive, and it cannot be set in motion sufficiently expeditious to meet sudden emergencies requiring that a person of unsound mind be immediately deprived of his liberty and placed under care and treatment. To meet the case, to secure proper protection for the lunatic, and also to ensure the termination of the disgraceful abuses which some years ago were practised in regard to lunatics, a standing commission was appointed under the title of the Lunacy Commission, the duty of which is to take cognisance both of lunatics and their keepers, and the Commissioners are required to visit periodically every person reported to be a lunatic, and ascertain that he or she is sufficiently and properly cared for.

Before a lunatic can be legally deprived of liberty or put under restraint, it is necessary that two certificates of insanity be made out and signed, each by a medical man, who must arrive at his judgment and conclusion separately from any other medical man, and upon personal examination of the patient.

I shall make it my business to tell you the law as far as it relates practically to the subject of Insanity, but at this point I lay particular stress upon the Medical certificates, since most of you will have more dealing with them than with the general details of lunacy law. You will find it useful now to pay particular attention to the general remarks upon certificates, as in your practice you will not always find it convenient to refer to books at the moment you want information. I regret to say that I have much too frequently found a disgraceful amount of ignorance regarding certificates and the manner of filling them up.

You may sometimes have a difficulty in obtaining access to a patient, but the difficulties often exist only in the imagination of the patient's friends. You will often be asked to employ stratagem to gain access, and if you follow the ideas of advisers you may often find yourself very much embarrassed. You are not a detective officer, and you will rarely fail to find a legitimate excuse for entering into the presence of the patient; and once in his presence you can in all ordinary cases ascertain the state of his mind without much difficulty. Some persons, however, especially those that have been under restraint before, are cunning and artful, and require some tact and management; but when you have once gained access your victory is half won, and you must pertinaciously adhere to your object of testing the mental condition of the patient until you have satisfied yourself as to his sanity or insanity.

I have been told by the friends of an alleged lunatic that the patient had sworn he would blow out the brains of any doctor who went near him; but I never have met with revolvers or blows on a first visit. The patient is usually thrown off his guard by the presence of a stranger and will commonly tell you his woes at once.

Although you are not called upon to subject yourself to personal risk, I believe that the danger of visiting a dangerous or artful lunatic is much less than is commonly imagined.

When requested to see a lunatic with a view of signing a certificate, you should remember that the duty, if you undertake it, is a very definite one. First, you have to ascertain whether the patient is insane; and secondly, you must determine whether he is a proper person to be placed under care and treatment. When you have examined a patient and concluded that he is insane and a proper person to be placed under care and treatment, you may proceed to make out a medical certificate of insanity which must be in the form prescribed in Schedule (A) No. 2. 16 and 17 Vic. c. 96; and in conformity with 25 and 26 Vic. c. 111, copies of which marked No. 1* I hand round to you, and they must be filled up in accordance with the instructions given in the Act and which you will always find printed with the form of the certificate. I need hardly remark that unless the certificate is filled up exactly in accordance with the legal form, it is invalid.

* All the forms referred to in these pages will be found in the appendix.

The form marked No. 2, which I hand round for your inspection, is a copy from an invalid certificate which was brought to me some time ago. As you will see, it runs thus:

"I, the undersigned, being a Doctor of Medicine and a Surgeon etc." but the act distinctly states that the qualification entitling a person to act as a Physician, Surgeon, or Apothecary must be set forth. The expression "Doctor of Medicine" does not set forth a qualification entitling the holder to practice, unless the University is specified, and in order that the qualification be available for the purposes of a lunacy certificate the University must be one, the degrees of which, like Cambridge or Oxford or London are admissible on the Medical Register. A Degree from the University of Giessen, for instance, would not be available for the purposes of a certificate.

Again, the expression "Surgeon" does not set forth a qualification. It is necessary, in order that the certificate may be valid, to state the College of which the Surgeon is a Fellow or a Member, as for example, a Fellow of the Royal College of Surgeons of England.

Form No. 3 is another example: the gentleman who signed it is a Fellow of the Royal College of Surgeons of England, but he omitted to specify "England" and thus his certificate was inaccurate. Scotland and Ireland both possess Colleges of Surgeons as well as England, and a certificate of insanity is imperfect if it does not specify of which particular College the person signing is a fellow or a member.

It is also required by 21 and 22 Vic. cap. 90 Section 37 that the person who signs the certificate shall be on the Medical Register, and as the Register only admits the qualifications recognized by the lunacy law, by a recent regulation, the expression " Registered Medical Practitioner" is now permitted to pass.

The next point in the certificate to be stated is the precise practice of the person signing it. The practitioner must not only be in *actual* practice but must be in practice as a Physician, Surgeon, or Apothecary.* To fill in the form as "being in actual practice as a general practitioner," as I have seen certificates filled, is a departure from the statute, and such a certificate if not rejected, might at all events be open to a question of legality if brought into dispute in a trial.

It is essential in order that the certificate may be valid, that the name of the street, and the number of the house (if any), or such like particulars of the house where the patient was examined be inserted. In the important decision of Mr. Justice Coleridge in the case of Greenwood, Feb. 12th, 1855 it was held that the certificate was invalid, because the name of the street was not stated.

It is further necessary that the patient's ordinary residence be stated in the place indicated for it in the form, and immediately afterwards the profession or occupation of the patient, if any : if the patient has no occupation the fact should be so stated, the form No. 4 which has the expression, "no occupation," may be taken for an example of

* 25 and 26 Vic. Cap. cxi. Sec. 47.

the words to be adopted. Form No. 5, has the expression "gentlewoman," and No. 6, has the wording "farmer's daughter:" these are samples of the expressions which may be used.

In form No. 6 you see an erasure and initials in the margin: you must remember if ever you make an erasure you must put your initials either against it or in the margin or your certificate will be rejected.

It is above all important that the patient's name be correctly written and that it agrees exactly with the name in the order and statement. Initials will not do; all the patients' names must be written in full, and each must be properly spelled, and agree in spelling with that of the order and statement. If the order and statement for example spell Mary Ann, your certificate must not be written Mary Anne or it will be invalid. Or, if the patient have more than two names you must write them all in full on every occasion thus: should his name be Henry Hubert Spencer, you must write Henry Hubert Spencer on every occasion, for the omission of a name or the substitution of an initial as, Henry Spencer or Henry H. Spencer will invalidate the certificate. It is necessary that in certifying, you describe the insane person either as a lunatic, an idiot, or a person of unsound mind.

The law does not recognise any other classification, and you must remember, whatever you think of your case, that you are drawing up a legal document, and you must implicitly obey its forms and not follow your own opinions.

To certify that A B is a person of weak intellect

will not provide a certificate upon which he can be deprived of liberty, the vague expression "unsound mind" is supposed to cover all the cases not idiotic or lunatic, and if the practitioner feels that he cannot conscientiously use either of the latter expressions he is bound to use the former: the certificate must state that A B is a lunatic, or an idiot, or a person of unsound mind.

It is absolutely necessary in order that the Certificate may have validity in law, that the medical practitioner who signs it shall set forth some fact or facts, or symptoms, indicating Insanity observed by himself. The strongest facts indicating Insanity are delusions and incoherence, and the statement of particular delusions or of the particulars of incoherence render a certificate most valuable. For example, if a carpenter should tell you that he believes himself to be a king, or the great God Almighty, or the Lord Jesus Christ, the *primâ facie* evidence is that of a delusion, and you may word your certificate, "he is labouring under the delusion that he is a king," or "that he is the great God Almighty, or the Lord Jesus Christ;" and if his conversation is incoherent, the statement of the fact that "he is incoherent" or that "his speech is incoherent," or "his conversation is incoherent," will in some cases strengthen your certificate; or you may word your certificate in this way, "whilst speaking to me the patient became incoherent, and unable to connect his sentences." Sometimes, however, the patient has not any delusions, as you may often observe in melancholia. In such cases your observation of the fact of the melancholy should be stated, and it should be strengthened by some other fact. The admission of a desire or of an attempt to commit suicide is strong confirmatory evidence of melancholia, and will furnish good grounds for forming an opinion of insanity, and signing a certificate; or the statement of the patient that he is depressed and melancholy, and that he feels himself at times under an impulse to commit violence which he cannot control, will justify you in filling up a certificate.

The following are samples taken from actual certificates which may be used in such cases: "he is suffering from melancholy, and informed me that he felt himself tempted to commit suicide;" or "he is suffering from melancholy, and informed me that in consequence of the great depression he was labouring under he had attempted to destroy himself;" or "he is the subject of acute melancholia, and informed me that he felt so depressed that he wished to destroy himself, and that at times he felt a strong, sudden, and overpowering impulse urging him to kill his wife."

Sometimes when you enter the room the patient will laugh but be unable to give you any explanation of his conduct; perhaps he will give expression to bursts of meaningless and unexplained laughter during the whole of your interview. Or the patient may be taciturn, and, either from pre-occupation of mind, from a cessation or paralysis of thought, or from loss of the faculty of language be unable or refuse to speak. Such facts would be highly valuable in confirmation of an expressed opinion of melancholia or dementia, but in themselves would be open to criticism unless conjoined with other evidences of insanity, as, for instance, refusal to take food, and neglect of the calls of nature. A person who is melancholy, taciturn, refuses to speak or answer questions, refuses to take food, and so neglects to attend to the calls of nature that he is wet and dirty, may properly be considered as of unsound mind, and a proper person to be taken charge of, and detained under care and treatment.

It is however all important that you make a careful personal examination, and elicit in that examination absolute facts indicating unsoundness of mind, and state them concisely and explicitly in your certificate, for a certificate may at any time be brought into a court of law, and if you have given a certificate upon imperfect evidence, or neglected to use due care, the case may be given against you, and you will render yourself liable to an action for false imprisonment. In a case tried some few years ago the judge remarked that the medical practitioner had not exercised due care in the examination of the patient, and the case was given against him accordingly.

The second part of a certificate, the statement of other facts (if any) indicating insanity, communicated by others, is not essential, though it is often useful as strengthening and confirming the first part, especially when it is desired to state the evidence of violence or the dangerous character of the lunatic, when the same has not been observed by the Medical Practitioner who signs the certificate, or when you feel it necessary to make clear the fact of a delusion, or of the neglect of food, or of a disregard of the calls of nature.

A gentleman told me the other day that he was ruined and had lost all he had in the world. The statement was a delusion, but *primâ facie* it might have been true. The gentleman, who had long ago retired from business, had a good income, and had not had the chance of losing any part of it, for his money was thoroughly secure in the hands of trustees. This information was imparted to me by one of the trustees, and its statement confirmed the fact of the delusion of ruin under which he was labouring.

It is however absolutely necessary that the name, both Christian and Surname, of the informant be given, otherwise the information will be useless.

It is not enough to use such expressions as father, mother, brother, sister, attendant, nurse or servant; the Christian and Surname of the relative, friend or servant must be given: as for example, if the name of a patient be William Scott and his friend's name be John Smith, and the friend informs you that the patient is violent and destructive, and it is desired to state this in the second part of the certificate, it should be expressed in some such form as the following:—John Smith, a friend of the patient, informs me that he is violent and destructive; or if some fact be expressed the evidence will be more valuable; as for example, John Smith, a friend of the patient, informs me that he is violent and destructive, and that yesterday the 19th inst. he put his fist through a pane of glass, and wounded a blood vessel on the back of his hand.

If a brother or sister, or brother-in-law, or any other relation give the information, it is best to give the Christian and Surname, and immediately afterwards the degree of relationship of the informant. It constantly happens that the general evidences of insanity are clear enough to demonstrate the fact to the practitioner's mind, though he is unable to elicit from the patient evidence sufficient in itself to warrant a certificate. It is then that the additional evidence afforded by a relative or servant who has been in constant attendance, comes to your assistance. The artifice of a lunatic is sometimes such that he or she will baffle every attempt to elicit facts indicating insanity, though the general phenomena of mental disease may be patent.

I was lately asked to sign a certificate of insanity against a lady, the wife of an officer in the army, who had been sent from India for the express purpose of receiving treatment in an English Asylum, and after an hour's conversation with her I found myself possessed of very little evidence that could be made available for a certificate. I questioned the patient closely on her family, her history, her stay in India, the people she met there, her mode of occupying her time, her health—in fact I questioned her upon every subject I could devise conversation upon, and all to no purpose, though her insanity was evident the moment she entered the room. It was discovered in a maniacal expression of face and manner—facts only to be learned by experience; it was also apparent in a readiness to communicate and confide her private affairs to me, though I was an absolute stranger to whom she had not even been introduced; and it was also evident in the expression of an irrational desire to engage and enter at once into possession of a furnished house, of which she had heard, notwithstanding the fact that she had never seen the house and that it was late on a Saturday night. This incongruity might however have borne a comparison with the customs of India, had I not learnt afterwards from the husband, who was present during the interview, that the suggestion of taking the house was a fixed idea, which no amount of reasoning would remove, and that if he had raised one syllable of opposition the patient would have felled him to the ground with a blow from her fist, though he was a very powerful man and stood six feet two in his stockings.

The facts, even with the husband's statement, would have afforded but slender evidence upon which to incarcerate the patient, and after a conversation with her husband in private, I renewed my examination of the lady. This second time I saw her alone, and I then elicited from her the facts of a delusion; she imagined that one of the surgeons of her husband's regiment wished to poison her, and that he had come from India at the same time that she did, and that she had seen him in a disguise among the steerage passengers in the steamer by which she came to Southampton.

The day after I saw her, she fled from the house of the friends with whom she was staying, and was not heard of until two days afterwards when she gave herself up to the police, by whom she was delivered to the authorities of the asylum for which she was destined. A week afterwards she exhibited such evidence of dementia that she did not know the day of the week, or the month of the year, or how long she had been in the asylum, whether days, weeks or months. I mention the case as a very striking instance of artifice and power of evasion sometimes exhibited by the insane, and I have found such artifice more commonly practised by those whose violence to themselves or others has rendered alienation absolutely necessary for their own safety, and the safety of those around.

Finally, I would remark that the expressions used in a certificate should be short and concise. The sentences should be logical and clear, and the composition should be studied and exact. I once saw the expression, "increase of morbid sensibility," stated as an evidence of insanity as though "morbid sensibility" was the condition of sanity, and its increase co-extensive with insanity.

The ridicule to which anyone exposes himself for illogical certificates is not undeserved, and if he, in consequence, finds himself considered as mad as the patient, he has only himself to thank for the wages of his carelessness.

On the table are a number of blank forms of certificates, and also of the order, and if you will, after the lecture, each take one, and at your leisure fill up the same with details of the examination of an imaginary case, the practice may be useful to you; and I shall have much pleasure in looking over and correcting the forms for you, if you will bring them on the occasion of our next lecture.

Persons Prohibited from Signing.

Certain persons are prohibited from signing certificates. Formerly the Medical Officers of Unions or Parishes were prohibited from signing in cases of pauper lunatics belonging thereto. This restriction has been withdrawn, but it is not lawful for a practitioner to sign a certificate against a relative, neither will a certificate be valid if the practitioner derive any profit out of the case beyond his honorarium. Neither are the certificates valid if the practitioners who have respectively signed them are assistants to one-another, or in partnership with each other, or in partnership with or nearly related to the patient, or to the person who signs the order for the reception of the patient. Neither can a practitioner sign a certificate for the reception of a patient into his own house, or into an asylum in which either he, or a partner, or father, brother, son, or assistant is directly interested. These points are points to bear in mind, and I would repeat, for it is important to remember, that the practitioner, however well he may be qualified, is prohibited from signing a certificate unless his name be on the Register.

In the case of pauper lunatics one medical certificate only is required, but in lieu of the second the patient has to be seen and the order has to be signed by a Magistrate, or by the Clergyman and Relieving Officer of the Parish, according to the form I submit for your inspection.

Unless, however, you are acting as the Medical Officer of a Union, or of a Parish, you will not be likely to be called upon to sign certificates for pauper lunatics. The ordinary rule is that pauper lunatics are removed to the Workhouse, where they spend one night at least, and whence they are removed to the County Asylum, upon the certificate of the Medical Officer of the Union; the order being signed by a magistrate, and the statement by the relieving officer.

The Order and Statement.

The order for reception and a statement of certain facts regarding the patient have to be made after a form prescribed in Schedule (A) No. 1, secs. 4 and 8 of 16 and 17 Vic. c. 96, copies of which I now hand you.

The patient must have been seen by the person signing the order within one month previous to the date of the order.

The directions given in the schedule of the Act and stated on the printed form you have in your hands are very plain, and yet from this example No. 7 you see how readily the form may be departed from in the face of plain directions.

As in the case of the certificates, so also in the case of the order it must not be signed by any person receiving any percentage, on or otherwise interested in the payments for the patient, nor by the medical attendant as defined in 8 and 9 Vic. cap. 100 sec. 90, of whom I shall speak directly.

Regulations for the Reception of Lunatics.

There is some ambiguity as to whether a person may receive a relative into his house or asylum for profit, and the ambiguity arises from the provision that the person who signs the order for reception shall not be the father, son, brother, partner, or assistant of the person who receives the patient.

The spirit of the Act however, clearly does not deny to people the care of their own relatives; but the interpretation of the word profit is a moot point and does not appear to have been satisfactorily settled. Great hardships might however occur were it insisted upon that every patient who requires to be sent away from home, must of necessity be sent to the care of strangers.

According to the printed statements I send round, you will observe that no person deriving profits from the charge, can receive into any house, or take care or charge of a patient, as a lunatic or alleged lunatic, without an order and two medical certificates. Should you at any time determine to take a lunatic into your house you must implicitly obey the following regulation.

Within one clear day after receiving a patient, true copies of the order and certificates, together with a statement (25 and 26 Vic. cap. 111. sec. 28) of the date of reception, and of the situation and designation of the house into which the patient has been received, as well as the Christian and Surname of the owner or occupier thereof, must be forwarded to the office of the Commissioners in Lunacy, 19 Whitehall Place, London, S.W.

The copy is best made out on one of the printed forms. Its first page bears the form of notice of reception. In it there is less to copy, and therefore less opportunity for error, and you must remember to be careful that the copy is a *true copy*. You should copy all erasures and initials of whatever nature or kind, for in this matter your duty is simply that of a clerk, and it is no part of your office to correct the mistakes of certificate writers.

If certificates contain errors the copies will be returned to you by the Commissioners, and the errors will be pointed out to you by them. You must then refer to the practitioner or practitioners who have made the errors, and require them to correct their errors and initial their corrections. You must then alter your copy according to the alterations of the original and return the document to the Commissioners. The limit of time for emendations is 14 days. You should therefore endeavour to obtain the corrections with all possible speed, for it sometimes happens that the correction is imperfect, and a certificate may be returned again and again. If the corrections are not satisfactorily made within the fortnight, the certificates expire, and the Commissioners will order the patient to be discharged.

In the event of such a discharge (provided the patient is still insane) you will require a new order and statement, and new certificates, copies of which must of course be sent, as before, to the Commissioners.

In the event of such new order and statement and certificates being required, there is one point you must remember, as it is essential; it is this—the patient must not be examined in your house. He or she must be removed and examined elsewhere, and when the new documents are duly signed, and not before, you may receive the lunatic back again.

Neglect of the regulations prescribed in the Lunacy Acts may make the offender the subject of a criminal trial, and the plea of not knowing the law will not be received as a valid and good excuse. It is the duty of every body who undertakes the charge of a lunatic to make himself acquainted with the law regarding lunatics before accepting the charge, and he will very properly be held responsible for his neglect should he fail to do so.

In the event of question afterwards arising as to the sanity of the patient, you will be held free from harm, provided you have fulfilled all the statutory requirements. An action will lie not against you, but against the persons who signed the order and certificates, and you may, in the event of an action, plead the certificates which with the order are quite sufficient to justify you in receiving the patient.

The law does not absolutely demand that every lunatic or person of unsound mind shall be placed under certificates. Thus a father or a mother is not bound by law to place a lunatic child under certificates, and there is no law prohibiting the treatment of a patient at home, provided of course that he is humanely treated, and that the father, or mother, or husband, or wife does not neglect him, and so bring himself or herself within the lash of the common law, for neglect or ill-treatment.

A friend even may receive a lunatic or an alleged lunatic into his house without certificates, provided he receives no profit from the charge. This is no doubt a wise law, though the liberty sometimes gives rise to great difficulty in dealing with a non-certificated case.

A person suspects that his relatives have deemed it right to place him under care and treatment, and he appeals to a friend and asks his friend's protection which is granted, and the patient is secure in his friend's custody, for who can enter another man's house against his wish?

The Commissioners' power does not extend to patients kept at home, or in the charge of friends who derive no profit from the care, and if a patient be in the house of a friend, even though against the wishes of his relations, a magistrate may refuse to issue an order unless some evidence of maltreatment is shown. In cases of alleged maltreatment, or alleged neglect by friends, if information be laid before a magistrate, or justice of the peace, he can visit the patient, or order the patient to be visited, and order his removal to an asylum.

The justice's power extends equally to wandering lunatics for whose care the magistrate can make an order, whether the lunatic's state be brought under his notice or whether he knows it of his own knowledge. This is a very wise provision of the law and it has released many a lunatic from continued and wilful cruelty. Cruelty, however, does not mean the restraint which is sometimes necessary to prevent a patient from doing harm to himself or others. If a patient be raving from any cause, whether from acute madness, or from delirium tremens, or from fever, and his safe custody is necessary for his own protection or the protection of those around him, it is perfectly lawful for any one to restrain him, no matter where he is, or what be the attendant circumstances.

You must, however, remember that in regard to the reception and custody of patients, the law draws its fast line at the derivation of profit; and it is your duty to bring the facts of the law under the notice of those who call upon you to attend insane patients, whenever you find that the provisions of the Statutes have not been fulfilled.

The Medical Attendant.

Every lunatic received as a single patient for profit, must be visited at least once a fortnight by a medical attendant.

. This medical attendant must be a physician,

surgeon, or apothecary, who did not sign either of the certificates, and who derives no profit, and who is not a partner, father, son, or brother of any person deriving profit, from the care or charge of the patient.

Thus you may be the medical attendant of a lunatic who is under the care of any person as a single patient, except your partner, father, son, brother or assistant; provided that you did not sign the order or either of the certificates, and provided also that you derive no profit from the care or charge of the patient.

This does not mean, however, that you are called upon to attend a lunatic for nothing. The labourer is worthy of his hire, and you can, of course, look to the person who engages you as medical attendant, for the payment of your fees. If you are the medical attendant of a single patient, you must visit that patient at least once in two weeks. You must, however, visit the patient after two clear days, and within one week of the reception, and draw up and forward to the office of the Commissioners (25 and 26 Vic. cap. 111, sec. 41) a statement of the condition of the patient according to the form in Schedule F 8 and 9 Vic. cap. 100, copies of which I also hand to you. The manner in which I usually fill up these documents is somewhat as follows: With respect to mental state he is the subject of mania or melancholia or dementia, as the case may be, and that with respect to bodily health and condition he is in good health and well nourished, or in feeble health and badly nourished, or he is weak and paralyzed, or any such particulars according as the facts may be. Another duty of the medical attendant is, at each visit, to make an entry in a book called the "Medical Visitation Book," according to form which I hand round to you.

This, as you see, includes a statement of the condition of the patient's health, both mental and bodily; and it is your duty to remark upon any change you may from time to time observe in either the one or the other. It is also your duty to notice whether the patient's clothing and person are kept clean, and you should note these in the book at your visit; you must also make observations upon the condition of the house, and comment upon them, particularly examining into the condition of the patient's bed, and the comfort and cleanliness of the apartments he occupies; and you should remember that a false entry in the medical visitation book is punishable as a misdemeanour.

The medical visitation book must be kept at the house where the patient is received, and with the order and certificates, where they are accessible to the Commissioners whenever they may visit the patient.

If you yourself receive a single patient you must engage a medical visitor for the patient; for failure in fulfilling this regulation will render you guilty of a misdemeanour.

Under special permission of the Commissioners the interval of visitation may be prolonged, provided the patient be placed to reside with a medical man. This is of course a matter of arrangement, to be made upon an application to the Commissioners; when however the permission is granted, a book, in exactly the same form as the "Medical Visitation Book," but called the "Medical Journal," is to be kept, and an entry must be made in it at least once in two weeks by the medical man who has charge of the patient. This book must be accessible to the Commissioners whenever they may visit the patient; and it is important that some person in the house besides yourself should know where you keep the certificates and books, in order that there may be no delay in producing them to the Commissioners, should they visit the patient at a time when you are not at home. Should the Commissioners at the time of their visit make any notes or write any reports in any of the books, the entries which they make must be copied out and transmitted by post to the Secretary of the Commissioners at their office without delay.

If the patient should die, notice of the death must be sent to the Commissioners and also to the Coroner of the district, and the notice must be according to the form you have before you.

In case of discharge, notice must be sent to the office of the Commissioners according to the form in your hands; should the patient escape, you must give notice of the escape, and also of the recapture should the same be effected.

If a patient escape you are at liberty to take

F

any measures, not in themselves illegal, for the recapture, which you may deem advisable. It sometimes happens that some of the relatives of the patient are averse to his being retained under certificates, and will, upon his escape, give him harbour or shelter. If, however, you find anybody harbouring or sheltering the lunatic you may without further ceremony exhibit the authority of your order and certificate, and take the patient; and, if necessary, and the case is one of danger or difficulty you may call any available aid to your assistance. If resistance is made on the part of the persons harbouring the lunatic you must of course proceed against them for illegal detention, or you may ask for the interference of a justice of the peace, or the aid of the police. You must bear in mind, however, that in the event of an escape the certificates expire in 14 days, at the end of which time your authority as regards recapture ceases, and the patient must be discharged.

It is important to remember that upon the request or demand of the person who signs the order for the patient's reception you are bound to discharge him, and any retention of the patient beyond the time specified in your order of discharge is illegal. If the person who signed the order for the reception should die, then the person who made the last payment in respect of the patient has the power of discharging him, but in making your return to the Commissioners you must state by whose authority the patient was discharged, as indicated in the form.

A patient may be removed from the care and charge of one person to that of another, the process, which is called transfer or removal, must be conducted under the consent and order of transfer of the Commissioners. The person who signed the order for the patient's reception is the proper person to apply to the Commissioners for the order of transfer, and the Commissioners, on satisfying themselves as to the propriety of the removal, grant the consent and furnish the necessary order in the form I send round to you. When the removal takes place notice thereof must be sent according to the regulation to the Commissioners, and on a form which the Commissioners usually supply. When a patient is removed from your care to that of another person, you are required, without charge, to furnish that person with a true copy of the order and certificates upon which you received the patient; you should also give a certificate to the effect that the patient is in a fit state of bodily health to be removed.

In the case of pauper lunatics transferred from one asylum to another, the orders of transfer are made by the magistrates instead of the Commissioners; but I need hardly occupy your time on this point.

You should make a rule of examining every patient the moment you admit him into your house, and record the fact of any marks or bruises which he may have, so as to shield yourself in case of after inquiry. If it be desired to give the patient liberty of absence anywhere for a definite time, the consent of the Commissioners must first be obtained. The application should be made to the commissioners in writing and they, after having satisfied themselves that there is no objection, will give their consent.

There is one point I must not omit to mention, viz. that every physician, surgeon or apothecary who visits a single patient, or under whose care a single patient may be, must, on the 10th of January or within seven days thereof in every year, report in writing to the Commissioners, the state of health, mental and bodily, of the patient, and such other circumstances as may be necessary to be communicated. You must be careful to remember this, it is easy to forget it, as it is an annual duty only, but if you become a medical visitor you should note it in your diary in order to have the matter at the right time before you.

LICENSES.

The subject of licenses is one we need consider but very briefly. You should, however, understand that no person is at liberty to take charge of more than one patient without a license granted by the magistrates of the county or borough, or if in the metropolitan districts by the Commissioners in lunacy. Should you desire to have your house licensed for the reception of more than one patient, formal application must be made to the magistrates, or to the Commissioners; you are required, at the time of making the application, to lodge therewith plans of the house you wish licensed, and to answer certain formal questions, after which your claims to a license are considered ; but a great variety of circumstances determine whether the permission is granted or not.

In regard to private asylums, lunatic hospitals, and pauper asylums, there are certain modifications of some of the regulations I have detailed. These you will readily learn should you become associated with any such institution, but they are not of such importance to you as practitioners that I need occupy your time with their discussion here.

In our next lecture we shall proceed to the consideration of acute mania.

LECTURE III.

Acute Mania.

Insanity generally—Definition of Insanity illustrative of Acute Mania—The term Mania, its definitions considered—Temperature—Pulse—Secretory and Excretory Organs—Excitement—Delusion—Cases of Acute Mania—Natural History—Prognosis—Photography in illustration of Mental Diseases—Etiology of Acute Mania—Pathology of Acute Mania—Imperfect recovery passing on to various states of Chronic Mania.

If you ask me to give you a definition of insanity I must answer, that definition is impossible, in the same sense as that definition of heat or of cold is impossible; insanity is antithetical to sanity and only known by comparison, as heat and cold are antithetical and only known by comparison.

We have not, however, in the case sanity and insanity, an arbitrary standard of comparison, as we have in the case of heat and cold; and even the ethical standard laid down by law, "the power of distinguishing right from wrong" is highly visionary, because in ethics there is no abstract idea of right and wrong. But as regards somatic phenomena, generally, we are much in the same position; we have no abstract idea of, and no definition of health and disease, except that they are antithetical, and only known by comparison, and we form for ourselves a sort of arbitrary standard of the healthy condition based upon harmony of action: thus the harmonious action of respiration,

Mania.

circulation, and the secretory, excretory, digestive, nervous, and locomotory organs, is called health, and the disturbance of this harmony constitutes disease, which may vary in every shade or degree, from very slight and transient disturbance to progressive and profound disorder. If the parallel be taken in mental conditions, and mental health or sanity be the harmony of the senses, the intellect, the emotions, and the will, insanity or mental disease will be the disturbance of this harmony: it may vary to every shade of degree, and like bodily disease, the pathology and pathological anatomy of mental disease will be referable to the arbitrary but scientific standard of a constant.

You will be able to gain some though not a perfect idea of the mental state in insanity, by a simple comparison.

The sane mind is able to correct an erroneous impression or observation of sense; a man waking in the night and seeing fiery writing on the wall is probably the subject of hallucination, and if he can correct this baseless creation of the fancy by an effort of reason, he may be considered to be sane; the same applies to false perception or illusion. But if a man imagines he sees an apparition in seeing a lamp post, he may well be considered as insane, if he is unable to correct the error by an effort of reason, and if either the illusion or the hallucination become fixed and permanent so as to modify the conduct of the patient, the mental affection must be considered as delusion, and the person must be considered as of unsound mind.

It is, however, my wish and my duty to simplify, if possible, a difficult study for you, and at all events to place before you the generally recognised, and most common forms of mental disease in a practical manner, and in such a way, that hereafter, you may readily recognise them for yourself. Without reference, therefore, at present, to any arrangement or classification more definite than that of acute and chronic, I propose at once to discuss with you the subject of mania.

You have seen three excellent examples of acute mania at Peckham House.

The first case, you saw on the occasion of our first visit, was a dark girl in a black dress with a red shawl over her shoulders, and who was thumping upon a piano and singing wildly, excitedly, and nonsensically, though in her sane state she was a good musician and sang well, and who spoke incoherently to us when we addressed her. Her excitement had been more intense two days before, when the foulness of her breath, and of the exhalations from her skin, would have offended your nose as much as her music offended your ears; and you remember when we saw her on the lawn last Friday she was almost well.

The second case, you saw, also on the occasion of our first visit, in bed; the patient at the time being exceedingly weak, with disturbed secretory and excretory organs, dry, brown, and cracked tongue, elevated temperature, and foul smell. On the last occasion upon which you saw her, you will remember she was up and dressed, but her hair was dishevelled and she screamed out loudly two or three times whilst we were in her room, threw her hands into the air, and then buried her face in her arms and in the covering of her bed, by the side of which she was sitting.

The third case, you saw on your last visit, was the girl who was lying in thick-quilted bed-clothing in the strong room, who used foul and abusive language, and declared that she had been told to use it by God. You will remember that you were told that she had torn all her clothing to shreds, and you could not have failed to notice the violence of her excitement, as indicated in her manner and speech and noise.

We shall now consider this condition.

Some alienest physicians would divide acute mania into two classes, and call one acute delirious mania, and the other acute mania. Dr. Blandford in his excellent work has drawn a line of division between certain cases of mania, but the distinction appears to me unnecessary. Some cases are much more acute than others, and the degree of vital depression is greater in some cases than in others. But this necessarily varies with the physique and constitution of the patient, and the variation which may be observed in the symptoms—by reason of the violence with which the disorder expends itself in some patients—does not appear to me to afford sufficient ground to constitute a variety upon. The pathological condition is the same in all cases of acute mania, and although it is a rapidly fatal pathology in some patients, and a rapidly recoverable one in others, the variation is to be found in the *vis vitæ* of the patients, and not in the pathology by which they are affected.

The term "mania" in its purely etymological sense means simply madness, but in our scientific nomenclature it represents a specific form of mental disease; it comprises a large class of mental maladies, and it embraces an almost infinite variety of conditions.

Mania has been defined as "delirium without fever, in which both judgment and memory are impaired, and the irritability of the body is diminished so as to resist many morbid causes." It has been treated of as a disorder of the reasoning faculties. Bacon, it is said, referred to mania when he declared that "madmen reasoned logically, but from false premises." It has been classed under the head of intellectual insanity, but no definition that has yet been attempted includes every condition in the multiple varieties and forms in which mania exhibits itself: neither is there any one pathognomonic sign of mania, unless it be that of undue excitation of one or more of the mental faculties. This indication, however, must not altogether be taken by itself, for although mania is always attended with undue excitement of one or more of the mental faculties, yet states of mental excitement, amounting to exasperation, may exist without the limits of health being over-reached. Undue excitement is, however, always attendant upon mania, and it is the feature you must endeavour to become familiar with.

Delirium, is certainly, not the characteristic of mania, though mania may sometimes be attended with delirium, and though fever is not always a prominent feature, yet in acute mania there is usually an increase of 2 or 3 or more degrees in the temperature. The common belief that maniacs are not as susceptible of morbid influences as sane people is based, I am convinced, upon some error in observation, and not upon statistics, and owes its probable origin to uncertain evidence similar to that from which the conclusion that a drunken man does not hurt himself when he falls is drawn. My own experience leads me to the belief that, not only as regards epidemic diseases, but perhaps all extraneous sources of disease, the maniac has no immunity by reason of his already pathological condition.

Disorders of the reasoning faculties, though a very common, and perhaps the most common, concomitant of acute mania, is not by any means an essential characteristic. Many cases come under observation in which there is neither disorder of the reasoning faculty, nor logical reasoning from false premises, nor any want of conception or memory, the mania being simply marked by a condition of excitement.

In the cases you have seen, and which I have just mentioned, excitement was a much more marked characteristic than either delirium or delusion; even in the case in the strong room, the declaration of the patient that she was ordered by God to use foul language, was more the outcome of violent excitement than of fixed delusion.

I lately saw a remarkably well-marked case of acute mania in which the reasoning faculties were unimpaired, though the excitement was undue and uncontrollable.

A young lady was brought to me who glided into the room with long, measured, sliding steps, and I learned that before she came into the house she had amused herself, and the public, by dancing in the street, but the most severe examination failed to discover any disorder of her reasoning faculties, indeed, her power of reasoning was acute, neither were there any signs of false premises in her logic, for she had no delusions. She admitted readily that her dancing in the streets was an incongruous act, and that the *à priori* conclusion regarding a lady who danced in the streets was madness, and that if, in consequence of her mad act, she was considered insane, she could not dissent from the opinion.

As regards her memory, instead of being defective, it was wonderfully clear, and, in calmer moments, she was able to describe the scenes of her varied travels with great accuracy and vividness. The memory of a maniac may, however, be much impaired.

In the undue excitation of some one faculty, as, for example, undue excitation of the emotional, the memory may be overwhelmed. But the impairment of one faculty is not necessarily the result of undue excitation of another: very often the impairment stands in the relation of cause to the excitation.

There may be, and usually is, a considerable disturbance and loss of the faculty of attention during the continuance of the attack, and Esquirol asserted that there is always a lesion of the attention in acute mania; and he was right.

Although imperfection of the faculty of attention is very commonly co-extensive with mania, it is not by any means conclusive evidence of an abnormal mental state, *e.g.*, in attempting a new study, it is often difficult to fix the attention, and, unless we can by an act of volition make an interest for ourselves in a new subject, we may readily wander into the varied fields of imagination created by relative suggestion.

It is undoubtedly true that in health we cannot control the sequence of ideas, but we can by an act of volition, place some new object before our minds, and so direct the current into another channel, and if we recognize, with Baillarger, that attention is "only the appreciation of the will," it follows that, in every mental state in which volition is interfered with, we may expect a loss, or an imperfection, in the faculty of attention. A person may be incapable, from an interference with volition, of placing any new object before his mind, and as he cannot control the sequence, many varied ideas may rise up in his consciousness in very rapid succession, and he may present any degree of incoherence, but this condition is not confined to mania, and may be associated with, or secondary to, any form of mental disease.

The lesion of attention is not necessarily attended with a loss. The maniac in his raving often seems to pay no attention, and yet is observant of everything, and will, after the excitement has passed away, repeat all that has been said in his presence.

It is necessary to remember this in practice, for you may have ordered medicine to be mixed with a patient's food, thinking that in his raging he would take no notice, and on the day following you will perhaps find that, in consequence of your order, he has refused food altogether.

Pinel observes, "He who has identified anger with fury, or transient mania, has expressed a view, the profound truth of which one feels disposed to admit the more one observes and compares a large number of cases of acute mania." "Such paroxysms, however, are rather composed of irascible emotions than any derangement of the understanding, or any whimsical singularity of the judgment;" but even this observation, which recognizes clearly the stage of undue excitation of emotions, is only dealing with that form of mania we crudely express as "raving madness," while any undue excitation may be considered as a mania, and the attendant disorders may be considered as consequences. For instance, we may say of the heart under abnormal conditions of increase in its action, that it is mad, or in a state of mania; or of the muscles in chorea, that their objectless contraction is a mania, or perhaps rather we should say that the mania was of the spinal cord, for in chorea there is clearly an undue excitation of the function of the cord. The same also may be said of tetanus, the contraction and spasm of the muscles being secondary. And with regard to mental phenomena, undue excitation of any mental function, dependent upon whatever cause, is a mania, and the incongruous attendant phenomena, influencing the conduct, are secondary and resultant.

The question, how does undue excitation of one function, or one set of functions, occur in acute mania? may here present itself to you. I think, however, the answer is more or less plain. Any defect or change in the brain material may, according to its seat, deprive that or some other portion of its normal control, when the potential energy of that uncontrolled portion will be rapidly expended, and hence the undue excitation of function. The harmony of the various functions of the brain is dependent upon the perfection of the whole organ, and the due and sequential performance of the various functions is dependent upon the control exercised by all working in harmony; if the seat of any one of these functions is disturbed, the performance of that function may become excited or depressed, but if so, it will not be astonishing if the remaining functions are disturbed or excited. An example may be found in other organs of the body.

The heart, for instance, often becomes excited and subject to palpitation when it is weakened by disease, but it is more commonly subject to excitement and palpitation when some other organ, as the stomach, is performing its work imperfectly: so with mental function, a loss of control over the seat of any function in the brain, whether the loss of control be primary or secondary, may result in excitement, or a liability to excitement, and this is the affection to which the name of mania is applied.

Perhaps the best short description, approaching to a definition, which I can give you of mania is, that it is—a state of mental excitement, associated with lesion of the attention, and causing actions which are incoherent and without rational motive.

Of course many objections may be made to this definition, but in every case of mania you will find in a greater or less degree, such a lesion of the attention that the patient either cannot fix it at all on anything, or else, that he has fixed it on some one thing and cannot fix it on anything else; and associated with this lesion of attention you will always find an uncontrollable excitement exhibited in manner, or speech, or both.

The popular notion that a maniac is incapable of understanding what is said to him, or that he is incapable of giving a rational answer, is not in accordance with fact; neither is it a fact, as is popularly supposed, that a maniac is a dangerous being, who can only be restrained by bars and bonds and locks.

Pinel, in 1792, was the first to demonstrate the fact that restraint in the treatment of mania was not necessary; and he was followed in this country by Dr. William Tuke, the humane founder of the

"Retreat", at York, who in the same year 1792 proposed the establishment of that Institution to his friends. Dr. Conolly, Dr. Gardner Hill, and numerous others followed in the wake of Pinel and Dr. Tuke, and the 'humane treatment' as it is called, of lunatics gradually became universal. Yet it is remarkable how, in the face of these well known facts, the popular notion of the dangerous character of the maniac should persist. The patients suffering from acute mania, whom you have seen at Peckham House, must have convinced you that there was nothing to be afraid of from them; and, as you will learn by and by, dangerous lunatics are rather to be found amongst certain epileptics, impulsive melancholics, and subjects of moral insanity, than amongst the miserable sufferers from acute mania, who are often more sinned against than themselves found sinning.

Having thus sketched an outline of mania in its primary and abstract sense, it will perhaps be easier to give you a clearer idea of the condition we have practically to deal with, by examining one or two cases.

J. B., a female aged 47, who was brought to St. Luke's Hospital during the period of my residence there, was said to have been in the condition in which I first saw her for 3 or 4 days before her admission; so excessive had been her excitement at her own home, that her friends deemed it necessary to convey her under an escort of three men and two women. Her certificates stated that she talked incoherently, and that she was under the delusion that all the solid food presented to her, was mixed with dust, and that in consequence she refused it; that she fancied that she went into a railway tunnel, and that all were killed except herself and a friend; she was also under the delusion that everyone was conspiring against her.

She was reported to have an utter detestation of food, and had hardly taken anything for days; she told me she was starving, and that she wanted some beef tea, which I immediately ordered, but which when brought she refused to touch. Her pulse was feeble and her skin hot; she was very thin and appeared very weak. Her aspect was wild and treacherous, and on taking her into the ward her action was as wild as her appearance. She immediately commenced to run riot and tried to injure herself by striking herself against the walls, or any object in her way. She was bathed, and put to bed almost immediately, and was again offered food which she threw on the floor; it was then apparent that it would be necessary to feed her. She roared and screamed frantically during the process of feeding, but was quiet for a short time afterwards. Subsequently, she sat up in the bed, and commenced beating the side of her bed, till her hands or rather fists were blue with bruises. She then became calmer and continued so till night-fall. During the night she was not only sleepless but occupied herself in beating the door and destroying her bed clothing; she was also wet and dirty, and first plaistered the floor of her room, and afterwards her hair with her solid excretion.

Notwithstanding the scrupulous cleanliness of the condition in which this woman was kept, being bathed as often as necessary, and her bed-clothing changed whenever she made them dirty, she had about her a smell which I cannot well describe to you, but one which is very characteristic, and one you will frequently find about patients labouring under acute mania. She exhibited a wild, fiendish expression, and whenever I went into her room she would catch hold of my coat and try to tear it, saying that I was her prisoner, and she constantly declaimed violently against her attendants for feeding her.

What the exciting cause of her extraordinary condition was, is not known. In her history there was nothing wherewith to trace any exciting cause, though the fact of her having had an attack of acute mania ten years before, was a sufficient indication of the existence of a predisposition. She was married, and in good circumstances, and as far as I was able to learn, had not been subjected to any depressing influences, such as family affliction or pecuniary worry, which is too frequently the exciting cause of madness, or rather I should say a cause of nervous exhaustion, the influence of which, a brain already weak, is unable to withstand.

The progress of the case was perhaps as astonishing as the origin of her malady. In ten days every maniacal symptom had vanished, and the patient, speaking with regard strictly to her mental state after that time, was well. She remembered all the circumstances that occurred before her seizure, but the period of her attack seemed to her a dream she could not realise. She became neat and cleanly in her habits, was calm, and free from all excitement, spoke rationally, and expressed herself as very grateful for all that had been done for her. But there were elements in her case upon which we may reason back deductively, and from which I think we may arrive at some conclusion as to cause. After the disappearance of the mania, the patient was exceedingly weak and her general health was bad, and there was evidence that her secretive organs generally had for some time previously been unhealthy and imperfect in their action, and that all the functions had consequently become impaired, and their action inharmonious. Her skin, her liver, her bowels, her menses were all irregular and imperfect, and their disturbance was, at least, an element in her insanity.

The points you may particularly notice from her case as characteristic, are in the first instance the excitement, which was uncontrollable and so great as to prompt to objectless violence; next in importance was her delusion and her incoherence, her depressed vitality, feeble and quick pulse, sleeplessness, disturbed secretions, physiognomy and smell. Practically the sleeplessness or asomnia is of the utmost importance, for unless it can be combatted the case is hopeless.

You will find asomnia or sleeplessness a very characteristic symptom of acute mania, and the concomitant feverish conditions and depraved vitality will be evident by all the ordinary expressions, and all the secreting and excreting organs will, in a greater or less degree, be disordered; the smell, already mentioned, due to chemical causes, is further evidence of depraved secretion, in which change and decomposition of the waste products eliminated by the skin rapidly take place. Some cases are particularly disgusting from the effluvium resulting from decomposition of excreta from the genital organs, particularly when sexual excitement is a prominent feature. In very acute cases I have very constantly observed furred tongue, with aphthæ or sordes on the lips and gums. Sometimes, and not unfrequently, I have observed the sloughs of the skin I mentioned in our first lecture, the epithelial layer becoming elevated, with a thin ichorous rapidly-decomposing and offensive fluid, which is constantly poured out, though the actual loss of substance in the true skin may not be great. Sometimes the vital depression is so great that the patient dies outright, and the possibility of such a consummation ought always to be looked for. I have seen this result in several instances, and when the excitement is very intense and the sleeplessness continuous, it is to be expected.

I mentioned, in passing, the physiognomy of the patient, and it may be well here to touch upon the point, as we often hear much said about the insane look or the insane appearance. The contortion of features and the furious expression of face presented by maniacs is the uncontrollable play of the histrionic muscles as brought into action by the excitement. The wild expression of the eyes so often commented upon is due not to the poetical idea of flashing fire, but to muscular action, particularly to action or varied contraction of the orbicularis palpebrarum. When the mania subsides, the natural condition of the features is usually restored. I have several times seen mania render hideous, the features of really pretty women whose natural appearance has been restored upon the subsidence of the mania.

Much stress has been laid on the features of the insane, but the study of physiognomy, though in its way of much value, is, as yet, far too embryonic to be of great scientific worth, especially in the study of mental disease. It has, however, occupied several able observers, and I may mention in particular, Dr. Diamond, of Twickenham, formerly Superintendent of the Surrey County Asylum, and now proprietor of Twickenham House, who has very successfully photographed many insane patients, some of which were described by the late Dr. Conolly in the Medical Times and Gazette. I am inclined to believe that bye and bye physiognomy will become of great use to us in the diagnoses of many diseases, and perhaps in particular in those which we term mental. Through the kindness of Dr. Diamond I am enabled to show you some of his valuable photographs, and I now hand round for your inspection portraits of maniacs the lineaments in which are depicted with such extraordinary

MANIA,
Electrical Condition of the Hair.

vividness and accuracy that they cannot fail to strike you.

In the excellent work lately published by Mr. Darwin on the expression of the emotions (a book you ought to read, if you have not read it already), are numerous and interesting observations on lunatics, several of them made especially for Mr. Darwin by Dr. Crichton Brown; and as of special interest, you may note his remarks on the hair standing on end. You will find a woodcut, Fig. 19, in Mr. Darwin's book, taken from a photograph of a woman with her hair in a marvellously erect condition, of whom Dr. Browne says "that the state of her hair is a sure and convenient criterion of her mental condition."

Mr. Darwin adds, that, "the extraordinary condition of the hair in the insane is due, not only to its erection, but to its dryness and harshness, consequent on the subcutaneous glands failing to act. Dr. Bucknill has said,* that a lunatic 'is a lunatic to his fingers ends,' he might have added, and often to the extremity of each particular hair." It is a common belief among alienists that lunatics do move their hair in the expression of their emotions; so that the idea of hair standing on end is not altogether a poetical one. I have often noted that the hair was bristly and crisp, but except under the influence of electricity I have never seen it stand up on end.

When patients are excited, you will find the greatest possible difficulty in making accurate

* Quoted by Dr. Maudsley, *Body and Mind*, 1870, p. 41.

clinical observations of temperature, or even of pulse and tongue; but when such observations can be made, it is right to make them, as they are the index of the patient's strength and general physical condition, and will aid you in your judgment as to treatment, and the special requirements of the patient. You will often find that patients suffering from very acute mania will get rapidly well; in fact the tendency to recovery in acute mania is far greater than you might expect, and it forms one of the most hopeful and encouraging features in lunacy practice.

You may, perhaps, obtain the most comprehensive idea of acute mania from the consideration of a simple uncomplicated case, such as a puerperal case, in which the pathology is comparatively clear, and the cause distinctly recognizable.

The following will illustrate:—A patient, M. W., æt. 40, was admitted into St. Luke's Hospital, five weeks after confinement with her first child, she having been married $3\frac{1}{2}$ years. She had always led an industrious and active life, had been fairly educated, happily married, and was in very comfortable circumstances, but her first pregnancy did not occur until she was nearly 40 years of age, and her labour was tedious and protracted, the liquor amnii escaped 34 hours before the birth of the child, and her strength was almost exhausted when the child was born.

Very soon after her confinement the lochial flow ceased, and her breasts ceased to secrete milk within a few days of the cessation of the lochial discharge.

Her exhaustion continued for a fortnight, when she gradually became the subject of outbursts of excitement, was restless and sleepless, and commenced talking incessantly and incoherently upon the most incongruous and absurd subjects. Within a few days she became so violent that she bruised and injured herself, and would have destroyed her child by throwing it into the fire had it not been taken from her. She had a distaste for food, and at the time of her admission into the hospital she obstinately refused to take any at all, and it was necessary to feed her; her violence was great, but she was easily persuaded. It was hardly possible to keep her in bed, and often on going into her room I found her standing in a corner, with a sheet over her head; but on being spoken to gently she would get into bed again. For some days she was very taciturn, and required not only feeding, but close watching, for she was constantly tearing her clothing, and she several times bruised herself, not only by direct violence, but by falling in her endeavours to get out of bed. She was constantly, not only wet and dirty, but filthy, and would smear her ordure upon every part of the room. Her bowels were very much confined, and it was necessary frequently to give her purgatives. But she was always more or less quiet after a complete evacuation thereby procured. After a fortnight had been thus passed she rather suddenly began to take her food well, and at the same time, she became clean in her habits, began to sleep well at night, and daily to improve, in fact, she gained strength, both bodily and mental, so rapidly that within five weeks of her admission she was reported to be recovered, and fit for discharge.

The pathology of such a case is extremely simple, being purely that of exhaustion, and exhaustion of brain in common with all the other organs, the condition or function of control being the first of the brain functions that failed; there was undue excitation of that portion wherein cerebration occurs, and the individual was not only unreasonable, but unreasoning.

As to her behaviour towards her child she gave signs of a ruling idea of her incapacity to feed her offspring, and hence the impression of impending danger which she sought to avert by slaying the infant.

You will notice in this case that the lochial discharge ceased, and that the lacteal secretion disappeared very shortly afterwards. Such is the natural history of most cases of puerperal insanity, and in a large number the commencement of the stage of recovery is marked by the appearance of some flux, sometimes by a mucous flux from the uterus, and sometimes by a return of the menstrual discharge. These points should be borne in mind, for in women in the puerperal condition, any disappearance of a normal discharge is a warning of danger. Sometimes the cessation of the discharge is sudden. In one of the most difficult cases I ever had under my care, the patient, who was very weak and delicate, suddenly lost her lacteal secretion nine weeks after confinement.

She understood her husband to say one evening, that he had lost his employment, and she afterwards described to me that she felt on a sudden, a pain in her breasts, but from that moment the lacteal secretion, which previously had been in abundance, disappeared; she commenced howling, and forthwith became maniacal, and continued so for 10 or 11 weeks, when her menses reappeared, and she rapidly recovered.

A remarkable case, also puerperal, was under my care at St. Luke's Hospital, and I mention it particularly because it contains a distinct hereditary history.

The mother of the patient was in the hospital, the subject of acute mania at the time of the patient's admission, and the patient herself had had two previous attacks. The predisposition to insanity was therefore demonstrated to proof.

The patient was admitted about twenty days after her confinement, having shown symptoms of excitement very soon after the birth of her child. On admission she wore a wild aspect, but told me she knew she was in St. Luke's Hospital, that she knew she was mad, but that she would get quite well; she spoke rationally, though an hour before, she had said to the doctor who signed one of her certificates, that "the devil was tearing her up." She told me that "she knew she was rational while speaking to me, but that at night time she would be excited," and she was as good as her word. She was restless, violent and noisy, and refused her food. She was very weak, and a few mornings after her admission, on asking her how she was, she said that "her head felt as though it was full of wool." Her secretions had been all wrong from the first. Her lochia had suddenly ceased, her lacteal secretion had never been properly established, her bowels were very much confined, and I found that their regular or irregular action had a very marked influence upon her mental symptoms. She was feverish, had a great distaste for food, a furred tongue and rapid pulse; she was very much excited when the bowels were unevacuated, and her stools were clay-like and wanting in bile. This condition was, however, recovered from under the use of an alterative (grey powder) and a mild purgative, and in about nine days she began to improve. She began to sleep well and to take her food well, her delusions entirely left her, and she rapidly regained her wonted strength, and was discharged recovered within seven weeks of her admission.

In her case evidence of the appearance of uterine or vaginal flux was not confirmed; but you may note the fact, that the commencement of the stage of recovery was marked by the restoration to normal action of the excretory organs generally.

Etiology of Acute Mania.

In the outline of etiology I gave you in our first lecture, I pointed out that causes were regarded under the heads of predisposing and exciting, and that there was good reason to believe that a predisposition to insanity was necessary before an exciting cause could bring about the phenomena of abnormal states of mind.

In all the three cases you saw at Peckham, there was, I was told, hereditary predisposition, and the exciting cause was clear. In the first patient the mania broke out suddenly, and was excited by a disappointment of love. She had once before suffered from a similar attack, which was complicated with so-called hysteria. The second patient had had several previous attacks, and the existing attack was excited by business anxiety, the patient having been the proprietress of a shop the cares of which were greater than she could bear. The third was a well-marked instance of insanity following child-birth. But in all these, you remember, we noted alarming vital depression: indeed, the cases furnished striking evidence of the excessively depraved condition of vitality, coincident with this disorder. The patients were severely ill, and ill of a disease involving the brain, and attended with extreme vital depression. The phenomena of acute mania are dependent upon abnormal, though not always demonstrable conditions of brain, and in all probability of the sympathetic system also.

In all the cases we have considered the evidence has clearly pointed to an existing predisposition, and then to exhaustion following upon the exciting cause. Imperfection or arrest of nutrition as the exciting cause of mania is often apparent, and exhaustion from want of food is often followed by determined abstinence from nourishment. I have seen cases so weak from inanition that they could hardly move, and yet obstinately resist the efforts made to feed them. Starvation is a fertile cause of acute mania. It is on the records of St. Luke's Hospital, that at a time before the County Asylums were generally established throughout the country, whenever the ribbon weavers at Coventry were thrown out of work, there were always numerous applications for admission for patients with acute mania in a starving condition from that district. Experiment has even suggested that the mind may be thrown off its balance by simple exhaustion of the brain in common with the other organs of the body. Some most interesting experiments were described by the Rev. Professor Haughton of Dublin, in his address, read before the British Medical Association at Oxford, in which the waste of the tissues was calculated upon confinement in a closed room, while fasting, but, as stated by Professor Haughton, it was impossible to complete the experiments on the human subject, as the mind had almost given way under the prolonged abstinence.

In an enquiry into the cause of acute mania among the patients in the City of London Asylum, where I was at one time the Assistant Medical Officer, and where through the kindness and courtesy of my friend Dr. Jepson I was constantly enabled to consult the case-books and make observations, I found that privation and want was the most common exciting cause, particularly amongst the poorer classes; continued semi-starvation—such as I observed in the histories of those who might be represented as factory hands, and the dependents upon trades bearing but small profits, as paper box makers, envelope makers, trimmings makers, waiting-room women, the wages of some of whom were barely sufficient to feed one person, to say nothing of a family,—constantly appeared as the excitant. In the next place I found anxiety and mental strain the exciting cause in a large number of instances; and thirdly, depraved vitality from constitutional disease.

An interesting case, in which anxiety and starvation together appeared as the exciting cause, was that of a Jewish Rabbi living in the East end of London, who ran into the street one night, absolutely naked, brandishing a poker in one hand, and a carving-knife in the other. No indication of insanity had been noticed in him previously. He spoke the Spanish language, and was unable to understand English, of which he could speak only a few words. When brought to the Asylum, he was very excited, and paced the ward up and down for days. He could not be prevailed upon to eat anything but bread, and even that he often refused, intimating that it was a fast day. After a week he became calmer, and I noticed that he was constantly crying, as if commencing to realize his condition. He was exceedingly grateful for all that was done for him, and with care and nourishment he soon began to mend, and all evidence of mania disappeared. The first improvement, however, was not permanent, and he relapsed, and though his friends brought him food specially prepared by members of his own persuasion, he would not always eat it, declaring that it was a fast day. The recurrence of the mania was, however, of short duration, and he made a good and a rapid recovery. As testified by his friends, he suffered from anxiety, in consequence of little debts; added to this, it is certain that he was predisposed to insanity; and there is no doubt, according to his history, that he was the subject of insufficient nourishment; and hence the fanning into flames of the smouldering maniacal fire, whilst the consequent confusion of memory as to his fast days, resulted in still further starvation and vital deterioration.

A case of very acute mania, dependent upon depraved vitality from constitutional disease, was that of a patient under my care in a private asylum. I was asked to see him in his mother's house before his removal to the asylum, and on entering the room I found him sitting close to the fireplace, from which position he did not attempt to move, but taking up the poker suggested that it was a good weapon to break anyone's head with, and asked me if I should like him to try if he could break mine. But he very quietly put down the poker, and I sat down beside him, when he asked me who I was. He then told me that he was the King of Wales, and that he was going that night into Wales to be married to a princess. He then requested me to ascertain the time the train would leave, and asked me to send a carriage to take him to the railway station. He was in a very weak state of health, refused his food, was sleepless and noisy, and spent many nights in shouting, and destroying the bed-clothing.

In this case no other exciting cause, than that of a depressed state of vital energy, was apparent. He was the proprietor of a prosperous business, but he was suffering from pulmonary disease, and his general, health had been failing for a considerable time.

Slight physical causes, such as a railway journey, sometimes seem to be enough to set up an acute attack in a predisposed subject.

I was asked to see a gentleman who became acutely maniacal whilst travelling. He left his lodgings at the sea-side apparently well, and on reaching London was taken into custody for making a disturbance at the railway station, and it then became evident that he was insane. The moment I spoke to him he commanded me to hold my tongue whilst he raised a prayer for the Princess Alice, he then threw himself on his knees and began to pray. He was wet and dirty, took his food badly, was sleepless and noisy, and died from exhaustion without at any time making the least improvement. In this case there was evidence of a predisposing cause, the gentleman had been invalided from India in consequence of sun-stroke. But the history of the exciting cause was imperfect, and it was not shown that he was insane before he started on the railway journey. When I saw him he was thoroughly exhausted, and his weak brain was exhibiting its defect in excited and imperfect performance of function.

It is of the utmost value and interest to study the causes of acute mania in every particular case, for upon your judgment will depend answers to questions of momentous importance to the families whose relatives you have to treat, and whose interests you must consider.

In a large majority of cases you will find hereditary predisposition. But it is not easy to ascertain the truth on this head, since the public consider that the admission of insanity is a slur on their family history. In cases with distinct history your expression of opinion should be guarded, for although the patient may recover the attack, he will be liable to a return of it, and what is worse, he may progenerate a race of beings all the members of which will be liable to develop insanity at some time in their lives. It is well to note here that a predisposition may be set up in an individual by accidental circumstances. I have seen cases in which a blow on the head has been followed by acute mania. One case, in particular, impressed me some years ago whilst going over St. Luke's Hospital: a male patient, who had been an officer in the French army, had the cicatrix of a large sabre wound over the right parietal region of the head, and he had become maniacal about two years after the infliction of the injury.

Fever, again, is a recognized cause, and, to the

public, often stands in the position of the straw at which the ideal drowning man will snatch. The relatives will admit that one member of a family has been insane, but in the hope that you will not deduce hereditary predisposition from the fact, they will assure you that the cause was fever. I must, however, guard you against such assurances, for they are delusive, and though I have seen several instances of mania following fever, I believe that as a cause, it more often stands in the position of exciting than of predisposing, if indeed it is ever a predisposing cause.

You will observe that, in all the cases of acute mania I have detailed, the history has pointed to exhaustion as the proximate and exciting cause; and I have endeavoured to show that a predisposing cause existed. I incline strongly to the view that a potentiality of insanity is necessary, and that insanity is not developed unless that potentiality exists, or, in other words, that insanity cannot be set up, unless in the patient there is a predisposing cause, and I make little doubt but that you will arrive at the same conclusion when you have made some observations for yourself, and will find also that, almost invariably, the immediate excitant of the attack is some exhausting influence.

PATHOLOGY OF ACUTE MANIA.

This is just the condition in which pathological anatomy tells us very little. Acute mania is a dynamical condition. The mental symptoms are evidence of profound cerebral disorder, but like the rubor, calor, tumor, and dolor of acute inflammation, it is transient, and leaves little or no trace behind it.

The most frequent *post-mortem* condition found in subjects who have died during an acute attack, is serous effusion; but this must be a result and not a cause, and may be coincident with the coma which usually supervenes shortly before death.

Sometimes considerable vascularity and injection of the vessels is observed, the brain even appearing to be swollen, as though it had been in a state of inflammation; but there is no evidence of inflammatory action having occurred. There are no products of inflammation. There is no lymph, nor even any greasiness of the membranes; there are none of the appearances of either cerebritis or meningitis, unless it be the congestion of brain and membranes, which is by no means constant, indeed, so unsatisfactory is the evidence furnished by the state of the vessels, and so liable is the brain to change its condition of vascularity after death, that it is almost impossible to say how much of the apparent congestion belongs to conditions which existed before death, and how much to the mode of dying, or to mechanical causes operating on the body after death.

Such coarse conditions as thickened membranes, adherent surfaces of the arachnoid, spicules of bone, tumours and growths, of course must be regarded as accidental. They, certainly, cannot be looked upon as the immediate exciting cause of the acute attack.

Professor Shrœder Van der Kolk described and gave cases of acute inflammation of the dura mater with mental disturbance, but I do not know of the record of any others, and I have never seen a case. It may be that pachyminengitis is a disease which is not very fatal. I shall speak of the subject again when we consider chronic mania.

Shrœder Van der Kolk also described an exudation of plastic lymph as occurring in acute mania, also a red inflammatory condition of the surface of the brain, and his observations have been copied into some of the more recent text-books. I very much fear, however, that the posthumous papers of this great observer may mislead. There are in the work compiled by Dr. F. A. Hartsen, of Utrecht, numerous valuable notes recorded by the late Professor; but there are also numbers of others which are only the passing sketches and thoughts which every author accumulates. To many of these Professor Shrœder Van der Kolk would never have given the authority of his name, had he lived to complete his work himself, and his note on the pathology of acute mania seems to be of this kind: his observation, nevertheless, in regard to the existence of cerebral congestion in association with acute mania probably is sound and correct.

In subjects who have died in the acute attack, most observers have recorded that they found the pia mater and sinuses full of blood, and the arachnoid opalescent; but you will see the same condition so frequently in the subjects examined in the *post-mortem* room of this hospital, that you will not be able to attach any importance to the fact.

Staining of the brain substance has also been recorded, but we cannot attach any importance to it, and I doubt if capillary apoplexies, which have been recorded in some cases, have anything more than an accidental presence. The more so because apoplexy, or as it should be called, cerebral hæmorrhage, occurs when the brain is anæmic rather than when it is congested.

Dr. Blandford* suggested the theory, "that the cause of delirium and death in acute and rapid cases of insanity, delirium tremens, and the like, is stasis of the capillary circulation, the result of pressure or inflammatory change in the blood." I confess that I have not been able to find any evidence of either the arrest of the capillary circulation or of the inflammatory change. Neither have I found the minute capillary vessels blocked with embolic masses in cases of acute mania. As mentioned by Dr. Blandford, Dr. Charlton Bastian† described some capillary emboli, which he found in the brain of an intemperate man, who died of erysipelas, with violent delirium, and he alluded to other cases of delirium in which the same state of vessels was observed. But that there is no necessary connection between the delirium of acute mania and the embolism, I think there can be no doubt, for the symptom is not constant.

* *Lectures*, p. 96.
† *British Medical Journal*, Jan. 1869.

Under one of the microscopes is a section of brain containing blocked capillaries resulting from embolism, but the patient suffered neither delirium nor chorea, which has also been said to depend upon capillary embolism. The patient who furnished this section became paralysed, and died comatose, his brain presented three red patches, the so-called acute red softening; in these the vessels were blocked with emboli, and the capillaries behind them retained their blood from stasis. But the man did not suffer from any mental symptoms. Under another microscope is a section from the brain of a person who died with some acutely maniacal symptoms, but in this section the vessels are empty.

What then is the pathology of acute mania? I think that we have ground to believe that it is cerebral congestion, and in support of this view we have some evidence. We may in the first place assume from an antithesis this much, viz., as sleep is brought about by a reduction of blood supply, so wakefulness will continue whilst the cerebral circulation is full.

In the next place, if we consider the action of stimuli, we shall find that they maintain a fulness of the vessels, whilst the stimulus of prolonged mental activity from voluntary effort will maintain an active fulness of the vessels, which beyond a certain limit will overcome sleep.

All mental activity is attended with increased blood supply, and the evidence afforded by the use of alcohol in quantities sufficient to stimulate but not to intoxicate, is that the increased supply of blood directly increases the mental activity. Van der Kolk used to bleed from the nose to relieve the congestion of the head through the ophthalmic artery, which distributes branches to the nose. In most cases of acute mania, the head is hot, though the patient does not complain of it, and doubtless he often has headache, though he does not observe it. Our evidence in acute mania points only to cerebral hyperæmia, and consequent mal-nutrition, which evidence is corroborated rather than negatived by observations upon the brains of those who have died whilst laboring under chronic mania. The latter often exhibiting the products of continued hyperæmia, of imperfect nutrition, and sometimes of the formative changes of a very slow sub-acute inflammatory action; but these we shall speak of in their place. It is for us here to note that the only certain evidence we have of the pathology is that of cerebral hyperæmia—itself a dynamical condition. Hyperæmia occuring in any other organ, as a gland, disturbs the function of that organ. It need not therefore surprise us to find that cerebral hyperæmia disturbs the functions of the brain, and that the disturbance should be earliest shown in the brain's most active function, namely, the mind.

We shall speak of the treatment of acute mania with the treatment of mania generally. But I may here observe that the salvation of the cases which recover rests upon their early treatment; and it should be remembered, not only in mania, but in insanity generally, that early treatment affords the best ground upon which to base favourable prognosis.

Imperfect Recovery Passing to Chronic Mania.

The group of cases we have considered, includes some of the most satisfactory we could have to deal with: in a large number of acute cases patients recover. It is true that many prove fatal, but from causes over which medicine can have no control.

Let us for a moment suppose one of the acute cases which we have considered to have been neglected, and not brought under the influence of treatment at all, nor under conditions favourable for recovery. What we may ask, would have been the probable issue of the case, if the patient had not died? There can be little doubt but that tissue alteration in the brain, resulting from its badly nourished condition would have taken place and become permanent, that the morbid excitability would have become chronic, and that the delusions would have become fixed and permanent also, though the general health might have become more or less restored. This in fact is what frequently occurs, and we find either a chronic state of excitement, usually moderate in degree, and very rarely violent, or else we observe one firm fixed delusion which guides and directs the conduct of the individual, an uncorrected impression, suggesting action, having become fixed in the patient's mind.

Sometimes the hallucinatory or illusionary impressions may for a time be more or less dormant, so that occasionally only, they give origin to delusions, and form premises upon which the person asserts the truth of some absurdity, which he is unable to correct by comparison, and upon the belief in which he acts, or allows his conduct to be governed.

A gentleman lately called upon me, and told me that at the house next door to the one in which he lived, some fowls had been poisoned, and that he believed that his neighbours accused him of poisoning them, and he believed himself that he had poisoned them. He, however, had never had any poison in his possession, yet he firmly believed in the truth of the circumstance, and in consequence he imagined that he was watched and followed whenever he went out, and he stated that he believed that as soon as parliament met, an act was to be passed, to condenm him without a trial. He told me his story calmly and coherently, and when I suggested to him that he was labouring under a delusion, he told me that he knew he was, but that he nevertheless believed it to be true. On another occasion, he told me that he was the subject of a conspiracy, and that the fowls had been poisoned by the conspirators, but that they had managed the matter so cleverly, that clear evidence would be brought against him, that he would certainly be condemned: he said he believed whenever he saw two people standing in the street talking, that they were talking about him. He said that he knew that he ought not to be at large, or that he ought not to be about alone, for he felt desirous of asserting himself, and interfering whenever he saw two people talking together. He however objected very strongly to my suggestion, that he should have some one always with him when he went out, as he said he was quite convinced, that any one whom he had with him, would very soon become associated with the people who were conspiring against him. This gentleman had had an attack of acute mania fifteen years before, excited in all probability by business anxieties. In that attack he laboured under the delusion, that he had committed some crime, and the morbid impression had never been quite effaced. At the time he came to me, his general health was a little below par, and the weak organ, his brain, made its imperfection known, by the exhibition of a little excitement, and revival of the old idea. I recommended him a change of scene, and advised him to go and pay a visit to a friend in the country for a time, and to live quietly and as much as possible in the open air. This he did, and he has since returned to town, much improved in his general health, and quite free from delusions.

A very characteristic case, of the chronic class, was under my care in St. Luke's Hospital. The subject, a lady of superior education, and of refined feelings and tastes, had an attack of acute mania some years ago, in which she became strongly impressed, that she was the "Woman of Babylon". I saw her whilst she was labouring under the acute attack; at that time, she was in a continual state of excitement. At night she would repeatedly call out, "whatever shall I do?" and in the day-time would constantly walk about exclaiming, "lost, lost, lost, and the sentence is for ever and ever." At the time of her admission, she would not listen to reasoning at all; but five years afterwards, when I took charge of St. Luke's Hospital, I found her, I cannot say perfectly rational, but able to converse pleasantly and rationally, upon almost any topic; the range of both her reading and of her travels was very great, and she took a passing interest in every event of the day, but on no consideration, would she leave her ward, nor would she, to use her own expression, face the world upon any inducement. She remained impressed as strongly as she had been five years before, with the belief that she was the Woman mentioned in the Book of Revelation, and any attempt to reason with her upon the subject, would have at once lighted up anew, a state of violent and uncontrollable excitement.

Sometimes chronic mania is manifested in a continued excitement, which though moderate in its degree, seems never to subside or intermit. What mind the patients have is constantly active. As a rule the subjects of this form of insanity are not violent, or are so only occasionally; in them, however, vicious propensities are often wonderfully exhibited. Sometimes the patients are loquacious and flippant, and incongruously merry; constantly they give expression to new delusions which take the place of old ones, and these frequently prompt cunning actions, or excite wantonness.

A patient we examined the other day at Peckham House was an excellent example of the persistent sub-acute or chronic mania. You remember the woman was loquacious to the utmost degree, chattering incessently about electricity with which she said she was affected,—of chloroform which she said had been thrown over her,—of voices which she said she heard speaking to her about insanity,— of a trumpet through which she said she heard one voice in particular speaking to her,—of a shoemaker's (or as she characteristically called it, a snob's) knife with which she said she had cut her arm, and of the various complaints which she made against persons, real or imaginary. All the sentences, however, were disjointed and disconnected and expressed without regard to logic or sequence, and delivered under the influence of a manifest though not excessive excitement. The hallucinations of sound and touch, under which she laboured continuously, put before her a new idea; and this new idea constantly, by relative suggestion, brought others into prominence in her mind. But from the loss of the faculty of attention, she was unable to keep any one of these ideas before her for a sufficient length of time to compare or correct it; hence the voluble jargon that must greatly have struck, if it did not amuse you.

A very characteristic feature in cases of this class is a propensity for mischievous stories and mischievous actions. Most of the patients exhibit an utter disregard of truth, always have a malicious story to tell against somebody, and often claim to be themselves heroes, and generally of marvellous and impossible exploits. By some the propensity for pilfering and hoarding is remarkably displayed.

LECTURE IV.

Chronic Mania.

Chronic Mania continued—Cases—Monomania So-called—Varieties of Mania — Epilepsy and Mania—Various Phases of Mind Associated with Epilepsy—Pathology of Epilepsy—Sleep-Talking and Sleep-walking—Catalepsy—Demonomanics.

In our last lecture we considered the subject of acute mania, and we commenced to pass in review some instances of the chronic form of the malady, and as an illustration from cases you have had an opportunity of observing, I may mention the woman you saw at Peckham House, on the occasion of our last visit, who constantly asked for kisses, and repeated the last sentences of every question ad-addressed to her. The curious repetition was perhaps accidental, and dependent on loss of memory in a great degree. Her mania is passing into dementia, and ere long she will have forgotten to ask for kisses, or how to repeat the nursery rhyme of *Little Jack Horner* with which she amused us. The case which you may have observed of a woman who declared herself to be various royal personages, and the Deity, and who on our first visit to Peckham House was offended because some of us did not make her the obeisance she considered her rank demanded, is as good an instance, perhaps, as you can see of chronic mania. When we entered the court in which she was, you may remember, she took no notice of us, but walked away apparently rapt in her own thoughts, and then came forward, called herself the Queen Charlotte, Julius Cæsar, Napoleon, Victoria, and Albert, Cleopatra the Maker of the heavens and the earth; declared that liberties had been taken with her, and then told us that she was *God*, that she could move the sun, and extending her arms tried to show us how she exercised her power.

The morbid fancy in this class of cases is often of long standing, but in some cases a new delusion crops up every day. The patients have a wild, weird expression of face, and, as a rule, are preoccupied with their delusions, and rarely take thought or care for any around them, except to abuse those in authority; and though they possess very fair, or perhaps even good, physical health, yet their mind is always warped, extravagant, and unreasoning, and if their excitement is not at the moment visible, a very short conversation will serve to exhibit it.

There was a woman in St. Luke's Hospital who had had three or four attacks of acute mania, and whose malady afterwards assumed the chronic form. She was perpetually noisy and semi-excited: she used to commence to talk loudly whenever any one entered the ward, and every day had some fresh, malicious, and unfounded, story to tell me regarding her attendant. One day she gave me a padlock she had stolen months before whilst the ward was being cleaned; the padlock had been searched for, but so cunningly had she managed to secrete it underneath the heavy fire-guard, that it was never found till she herself produced it. She was unable to sustain a conversation for long, and became incoherent soon after commencing to speak, but there was, withal, considerable method in her madness, and she seemed to take a fiendish delight, when she could not sleep herself, in keeping the other patients awake by making a noise; and she systematically, and night after night, befouled her room with her excrement, rendering it almost sickening to inspect.

There was another case of great interest at St. Luke's: J. A., formerly a clergyman of the Church of England, who was the subject of acute mania which became chronic. He used to chatter all day to himself, and often talked all through the night, would answer rationally if spoken to, but at once recommenced his chatter after giving his answer. He was very easily excited, was very pugnacious, and frequently used to deal violent blows without the least provocation.

You may have ample opportunities of observing similar cases at Peckham House, and if you visit County Asylums you will certainly see plenty of cases of chronic mania. You will find the patients comparatively well in general health, and, very frequently, employed in the various branches of trade and industry provided for their occupation, and yet the constant mumbling of something incoherent, or a frequent, perhaps almost fiendish, laugh, will demonstrate the preoccupation of their minds, and the low form of excitement disturbing its equable balance. In the

City of London Asylum, I remember a remarkable case. The patient was a shoemaker, he was a very good workman; though he had been insane for years, he always wore an excited expression, and always believed that he was the subject of a conspiracy. He had suffered from an attack of acute mania, during which this idea had formed. His general health had recovered, but not before so much change had occurred in his brain that the delusion had become fixed and unalterable.

From what I have already said of chronic mania, you must not by any means think that it only occurs as a sequel to acute. I have placed chronic mania by the side of acute to show it in its strongest light, and to reason back to its pathological condition, as far as we possibly can, from the materials we possess.

But we may consider it as dependent upon change more or less permanent in the surface of the hemispheres, whereby control becomes always more or less imperfect, whilst the normal and ordinary function of comparison is more or less in abeyance, being overwhelmed by the excess of the subjective phenomena resulting from the state of excitement.

Although many cases of chronic mania are the sequel of acute, yet very many are essentially chronic from their commencement, and, according to Griesinger, many cases follow acute melancholia. He mentions one exceedingly well-marked case in point. The patient, K——, formerly lively and sociable, became gradually, for about a year, meditative, taciturn, irritable, and solitary; he often used secret remedies, and always showed distrust of those by whom he was surrounded. At last he openly declared, " I feel myself very unwell, I have within me a putrifying mass which destroys my inwards; my neighbours therefore treat me with mockery and contempt, and avoid coming near me because I emit a pestilential odour." He led a solitary and sorrowful life; his delirium became always more confirmed, and he accounted for his disease by infection from glanders. He removed to a strange town, and took a walk to see whether those whom he met would also avoid him on account of the bad smell. As by chance a passer-by put his pocket-handkerchief to his nose, and at the same time looked at him, K——, violently attacked him and called him a hard-hearted mocker, an uncharitable despiser of men, and gave him a box on the ear. He was then recognized to be insane. It was found that he was insensible to external odours; he declared that he felt only his own smell, which resembled that of horse's urine, and he complained also of a corresponding taste in his mouth.

It is in this class of cases, essentially chronic from their origin, that we often find examples of the form of insanity commonly spoken of as monomania. I suppose it is hardly necessary for me to tell you that there is no such form of mental disease as monomania. A monomaniac is a person with a delusion and essentially *mad*, and though his delusion may for a time consist of only one fixed abortive idea, yet that, influencing his conduct and regulating his course of life, may impel him at any time to the commission of any unreasonable act; therefore, although he only exhibits one delusion, he is unquestionably a person of unsound mind. The subject of delusional insanity will enter considerably into our discussion of melancholia, so that I shall reserve most of what I have to say on the subject until we undertake that consideration.

It is, however, not uncommon to find chronic maniacs the victims of a fixed delusion, which they will conceal often with extraordinary skill, though they are ever on the alert to fulfil the promptings of their morbid fancy.

In illustration of this class, I may mention the case of J. B., formerly in St. Luke's Hospital, who had never had an attack of acute mania. His general health was good, and his conversation rational on most subjects, and the ordinary observer would never have thought him mad. He had been a clerk in a government office, and assumed the idea that the authorities had conspired to injure him, and he vowed vengence against them; but he never spoke of the subject unless it was mentioned to him, when he declared he would carry his threat into effect the moment he set foot outside the hospital. I was never able to allow him the privilege of walking in the streets, as he had expressed to me his intention of running away, and setting fire to the public office in which he had been employed, if ever he got the chance.

W. F. was another case, to all appearance sane. The man could converse calmly and rationally on almost every subject except politics, upon which, however, he used to become greatly excited, and would declaim and gesticulate violently, because he was not a member of parliament. But he never spoke upon political subjects unless some one introduced them to him.

Varieties of Mania.

The so-called varieties of mania, we may best consider here. Some of them will merely require mentioning, others must be considered in detail. One of the most important is that associated with epilepsy. The insanity associated with epilepsy frequently, though not always, takes the form of mania. Pathologically, the insanity which complicates epilepsy does not constitute a distinct class, but it is often regarded in the light of a distinct class by systematic writers.

The manifestation of the combined disorders, epilepsy and insanity, vary very much. Sometimes the subjects are very passive, and pass as it were through a sort of trance, sometimes the mental disturbance will be shown in stupidity and loss of memory, sometimes the patients will have a fit of greater or less intensity and for some time afterwards remain in a state varying from a slight degree of excitement to that of demoniacal violence, and sometimes they will have an insane paroxysm which will last for some hours and terminate with a fit. But the remarkable fact connected with the abnormal phase of mind often attendant upon the epileptic fit is, that on recovery the patients remember nothing of what has occurred during the time the effect of the fit lasted, or if they have any memory of it, it is only that of a hazy fading impression resembling the remembrance of a dream.

There are few subjects, perhaps, of more importance in their moral, and also in their medico-legal aspect, than the insanity associated with epilepsy, for if you can demonstrate to proof that a criminal act was committed under the influence of the epileptic disturbance of mind, the accused will not be held responsible for the act.

M. Jules Falret, one of the ablest observers and writers on this subject, has stated that all epileptics are irresponsible beings, and though I cannot go so far as to say *all*, yet the more I see of epilepsy the more inclined am I to believe that the subjects of that appalling malady are not responsible for many of their acts.

I may illustrate the subject with some particulars of the case of cut-throat I mentioned in our first lecture. The patient, a girl who was eighteen years of age, was found one day by her mother lying insensible on her bed with her throat cut. The girl was brought to this (Guy's) Hospital and the cut rapidly healed. After the infliction of the wound the patient was insensible for some time, but when consciousness returned she was as shocked and horrified at finding her throat cut as was her mother on discovering the sad calamity which had befallen her daughter. The girl in fact knew nothing of it—knew nothing of having done it, and instead of desiring to die, her greatest anxiety was to live. Her friends did not know that she was epileptic, for she had no symptoms which were sufficiently patent to be recognised by those around her. But on pressing my inquiries, her mother told me that the girl had had a fall and struck her head when a young child, and that she was subject to fits of fainting, which I at once suspected to be fits of *le petit mal*, and my suspicions were afterwards confirmed by two well-marked seizures, which were attended with very distinct mental aberration lasting for some considerable time, added to which there were other confirmatory symptoms of epilepsy, viz. involuntary micturition, petechiæ on the face and chest, and temporary hemiplegia lasting from twenty-four to forty-eight hours. On each occasion she wetted her bed unconsciously, and I would ask you to take particular note of this fact, as a wetted bed is very often the first clue you get to the existence of epilepsy. A characteristic of epilepsy, and one of great importance in the diagnosis of obscure cases, is the production of petechiæ or subcutaneous ecchymoses which on careful examination you will find on the forehead, throat and chest, and also, if you examine with an ophthalmoscope, on the retina; it is a point I would draw your special attention to, as it is often an unobserved but significant evidence of epilepsy.

If a patient be presented to you with ecchymoses of face, neck and chest developed in the night varying in size from a flea-bite to that of patches the size of a sixpence, you may be almost certain that he has had a nocturnal seizure, especially if he complains of headache; should a patient complain of headache and of having unconsciously wetted himself, or of having involuntarily passed a motion, you may almost affirm that he has had a seizure, but should he in addition exhibit petechiæ you may affirm with absolute certainty that he has had a fit. If too a patient exhibit occasional insane symptoms, recurring suddenly, and with them petechiæ, you may affirm that he is epileptic.

I would lay stress too upon the slight and transient character of the somatic symptoms in some epileptic attacks, and especially those which often induce an abnormal state of mind. And it is important to remember that the phase of mental disturbance associated with epilepsy is also frequently fleeting, otherwise you may find great difficulty in proving your case to be epileptic. I have seen a patient sit in a state of semi-insensibility for a few minutes, then start up and declare that there were wheels of fire in the air. In this case the state of excitement and hallucination usually lasted half an hour, after which the patient used to fall asleep: but on waking all his mental disturbance had passed away, and he knew nothing of either his excitement or his fit except that he had a vague impression of having had a dream.

Dr. Jules Falret divides the mental maladies of epileptics into three classes.

1st. "Those in which the mental disturbance manifests itself in the intervals of the epilepsy and may be independent of it."

2nd. "Those in which the mental disorder occurs as a temporary phenomenon, and either precedes, accompanies, or follows the epilepsy."

3rd. "Those in which the mental symptoms occur in paroxysms of greater or less duration, either directly with, or independently of the vertiginous or convulsive phenomena."

Dr. Falret's classification is, however, rather discursive, and I think we shall find that the division may be resolved under two heads.

1st. That in which the epilepsy and the insanity are more less independent of one another.

2nd. That in which they are so associated as to be more or less dependent upon one another.

The first class of case is very common in asylums. The constant and habitual state of the patients is insanity, from which they appear never to be entirely free, their epilepsy recurring at greater or less intervals, and seemingly independent of their insanity. Sometimes the patients drop down, pass through one or more fits, and recover without apparent alteration in their mental condition, sometimes they exhibit an increase of their habitual excitement, but the tendency of all these cases is downward, and towards absolute and complete dementia

In the relation of cause to effect I should place

the insanity before the epilepsy in this class. We may say we have insanity because we have an abnormal brain, and as a second pathology we have epilepsy, the consequence of conditions in the skull, whereby the sudden contraction of the cerebral vessels and surface anæmia, which is the immediate cause of epilepsy is produced.

In the second class of case I should regard the conditions in the skull, whereby the sudden contraction of the vessels, and the surface anæmia are produced as the primary pathology. The mental phenomena being secondary, or the result of temporary changes induced in the brain by the same dynamical conditions which give rise to the epilepsy. Under this second class some remarkable mental states occur.

Trousseau,[*] who was deservedly eminent as a shrewd observer, remarked among the fleeting phenomena attendant upon epilepsy, often a singular changeableness of feeling, of temper, and of character, violent fits of passion, which the persons cannot master, and he regarded these as pointing to mental conditions which will be followed by phenomena more distinctly characteristic of cerebral disorders.

In some cases, the epilepsy is a mere giddiness or vertigo, and in some cases the giddiness even is not apparent to the observer, and all that is observable is the transient mental disturbance. Sometimes in incipient cases, the disturbance will be shown in a perverted speech. A case is related by

[*] *Clinical Lectures.* Translated by P. Victor Bazire, M.D.

Dr. Forbes Winslow of a lady who, at times, in attempting to repeat the Lord's prayer gave utterance to "Our Father which art in Hell," instead of "Our Father which art in Heaven." I have once or twice met with cases in which the patient so misplaced words as to use hot for cold, wet for dry, and similar misappropriations and reversions of simple language. Sometimes I have seen temporary aphasia remarkably well marked. One patient, a girl, at one time under my care in the Infirmary for Epilepsy, used, sometimes, to be unable to articulate more than the word "yes," and a man, who was under my care at about the same time, used to suffer from attacks lasting about two hours, during which he was absolutely unable to articulate, even though he was conscious. during the greater part of the time, of what was passing around him.

Sometimes the conduct of epileptics undergoes strange vicissitudes in very short times, passing from gay to grave or grave to gay, they may become, without apparent cause, peevish, desponding, depressed or sad, or they may become irritable, magnify little worries into annoyances of great magnitude, engage in undertakings of rashness and hazard incompatible with sound judgment, or lend themselves to quarrels and to violence, which have no seeming origin and which are inexplicable unless the epilepsy be recognized. Sometimes the patients suffer from confusion of ideas and loss of memory, sometimes their memory was never so powerful or their intellect so brilliant. But whatever marked the stage of excitation during the continuance of the disturbance, is as a rule forgotten entirely when the brain recovers itself. Sometimes they have a memory of what has passed, a memory vague and like that of a dream, usually very painful and depressing, or a feeling as though something had occurred, but a something they cannot remember or cannot account for.

Jules Falret, remarking upon the actions, and obvious phenomena presented by patients in this somnambulist-like condition, says " no one can form an accurate notion of the irascible feeling which suddenly seizes the epileptic and urges him to strike or smash anything near him.

During these transient attacks of furor, he is dangerous to those around him as well as to himself, and the attention of medical men cannot be too earnestly drawn to these conditions of instinctive and blind violence, which all authors have pointed out as frequent results of epileptic fits.

They may lead to the infliction of grave wounds, to the commission of suicide, of homicide, and arson, and yet the individual cannot be held responsible for the acts of violence perpetrated by him during this perfectly automatic though short-lived delirium."*

A patient who was under my care was one day seized with an attack whilst I was standing by him : so momentary was the attack that it was hardly observable, but he instantly rushed to the table

* *Archives Générale de Médecine*, Dec. 1860. Quoted by Trousseau.

and drew from it a chair, upon which another patient was about to sit down.

On another occasion I saw him fill his pocket-handkerchief with stones, intending, as he said, therewith to kill his attendant. He was however habitually insane, the subject of chronic mania passing into dementia. His history was rather curious: he had been a man not only of ability, but of note, his insanity appeared suddenly and became chronic, after which his epilepsy appeared. His mania was never great except on the recurrence of a fit, when he used to become very violent.

Another case, at one time under my care, was that of a young woman who during the intervals of attack would more or less recover her reason, but during the fit was wild and furious to the utmost and constantly endeavoured to commit suicide by striking her head against the wall or against the floor; it was impossible to treat her anywhere but in an asylum, as her attacks were very frequent, and her violence in them so great, that she required the safety of a padded room.

A third case was that of a girl of a naturally gentle disposition who became the subject of occasional attacks, in which she endeavoured to injure those around her; after the attacks had passed away she knew nothing of her conduct during its continuance.

I saw a gentleman the other day, of whom I learned that on two occasions he had suddenly risen from the dinner table, walked to a corner of the room, made water, and then returned to his seat quite unconscious of what he had done. On another occasion he lost his way completely in the neighbourhood where he lived, and was so utterly unable to give any account of himself that he was taken into custody by the police, and kept in their charge until his friends were found. On a third occasion, his right arm was found to have become paralysed during the night; the paralysis was very transient and soon passed away, but it pointed to an origin like epilepsy, and the patient I learnt was subject to attacks of violence and excitement of which he remembered nothing when the paroxysm had passed. His friends had never seen him in a convulsive fit, and therefore doubted the epileptic character of his malady. I however had but little doubt of it, and my opinion is supported by the fact that the gentleman is becoming gradually, but certainly progressively demented.

I would here ask you to note, in passing, that convulsion of the physical frame is by no means an essential characteristic of epilepsy. Convulsion is rather to be considered one of the many epiphenomena of epilepsy, which may be absent.

The pathognomonic sign of epilepsy is loss of consciousness, and the condition of the brain during the seizure is that of anæmia.

I may mention that, following the experiments of Drs. Brown-Sequard, Schrœder Van der Kolk, Kussmaul and Tenner, and others, I instituted a series of experiments myself from which I drew the following conclusions.

1. Epilepsy is a contraction of the cerebral ca-

pillaries and small arterial vessels: the order of its stages are:

α Irritation of the brain, either primary and direct, or secondary, and resulting from exhaustion.

β Contraction of the cerebral capillaries and small arterial vessels.

γ Cerebral anæmia and consequent loss of consciousness.

2. The muscular contraction and spasm, together with all the varying phenomena associated with epilepsy, are altogether secondary, and not at all essential or constant, but they are all manifestations of imperfect nervous control, or a loss of balance between the nervous and other systems.

The experiments and observations were made upon guinea pigs and rabbits, which I rendered epileptic by dividing a lateral half of the spinal cord transversely.

It would occupy too much of our time now, to detail these experiments to you, but their general results coincided with the observations of Brown-Sequard and Schrœder Van der Kolk, and also with those of Kussmaul and Tenner, who have worked upon the subject; also those of C. Westphal, of Berlin, who has lately repeated the experiments.

The conclusions too, which I have drawn from my experiments, are confirmed in a greater or less degree by the clinical observations recorded by Trousseau, Falret, and Brown-Sequard. In effect the clinical phenomena of epilepsy are, 1st. the appearance of pallor, accompanied by loss of consciousness, of greater or less duration, and, 2nd.

the flushing of the face, and the muscular contraction and spasm.

The pallor is due to anæmia, and is co-extensive with, and part of the anæmia which occurs within the skull.

The flushing of the face, the muscular contraction, and spasm or convulsion, and all the other phenomena, are secondary and consequent. But what I wish you particularly to note, is the primary and essential condition, viz., the anæmia and loss of consciousness: and also, that this anæmia and loss of consciousness may be so transient, that the patients may not even fall, they may feel momentarily giddy and the attack will be over. Sometimes the patient will effect what is called fainting, but whenever you have complaints of fainting you will do well to examine them and see if they are not evidences of epilepsy. Further, whenever your patients complain of vertigo, the first question you should ask yourselves, is whether or not the case before you is epileptic or whether its phenomena are not epileptiform, and after duly considering the evidences presented to you, I am sure that I am not saying too much in affirming, that in almost all cases of both vertigo and fainting, if you give your decision in favour of an epileptic origin of these phenomena, your diagnosis will not be wrong.

I mentioned a case in which a gentleman left the table, and made water in the corner of the room, and then returned to his seat. Trousseau mentions a similar case which happened in his practice. The patient, a French judge, got up from his seat one day whilst the court was sitting, muttered some unintelligible words, went to the council room, and returned a few seconds afterwards unconscious of having left his seat. When his colleagues asked him where he had been to, he assured them that he did not recollect having moved from his place. Shortly afterwards he got up in the same manner and the usher was told to follow him, and he was seen to enter the council room, make water in a corner and then again return into the court, but he was altogether ignorant of what he had done.

The same gentleman one night left a meeting at which he was discussing some historical questions, ran, without his hat, into the open square, avoiding the carriages and passers by, but recovering himself returned to the meeting.

Sometimes whilst reading a book he would suddenly cease, and would then repeat with volubility the last portion of the phrase at which he had stopped. His physiognomy wore an unusual expression at such times, but he almost immediately took up his book again and resumed his reading.

I might enlarge this part of our subject by the details of almost numberless cases, some of which have come under my own observation, and others which have been recorded by writers celebrated for their accuracy and power of observation; but further illustration is unnecessary if I have succeeded in making plain to your minds, the fact that the epileptic phenomenon sometimes occurs and continues for only so short a time as a moment, or even less, but in that moment, may so disturb the sensorium, that the patients, for a greater or less time afterwards, are wanting in self-consciousness, and though they may have a certain amount of apparent consciousness, they are not under the control of their ordinary mind, and are not known to themselves.

This is perhaps a little difficult to grasp at first, but it is nevertheless a fact.

Whilst the acts of persons in this condition are automatic, they are, more absolutely than in any other state, beyond the person's responsibility. The somewhat analogous phenomena with which you can compare such acts, are those of sleep-talking and sleep-walking, and the performing of certain feats, such as dressing during sleep, which a person may accomplish automatically, and remember nothing of upon waking. A case was related to me of a coal-miner, who used to dress, leave his home and go to his work in a somnolent condition, and who met with his death, whilst in this state, by falling into the basket in which he was about to descend the shaft. The most remarkable case that has ever come under my own observation, is that of a lady who, on two or three occasions, asked for and drank cold water during her sleep, but upon waking has been ignorant of the fact.

The unconscious action of epilepsy resembles this state, and the subjects certainly are not responsible for their acts.

Among the cases related by Trousseau, is one of a young man who fell down suddenly, but soon got up and commenced striking the passers by with violence. He was taken to the police station where he for some time continued to insult the soldiers who held him, and spat in their faces.

It is easy to imagine the unenviable position of anyone charged with such an offence if he had no witness of his vertigo or epilepsy, or no physician to demonstrate his mental condition, and his consequent irresponsibility to a magistrate or a jury.

I ought not to omit to mention the condition called catalepsy, as an analogous state, and a further illustration.

Catalepsy is really the same disease pathologically as epilepsy, but possessing a few characteristics by which the older observers were led to separate it into a distinct class.

In catalepsy the patients' self-consciousness is in abeyance, for they remember nothing that occurs during the time they are under the influence of the seizure; and though they may speak or sing during the continuance of the attack, they are so unconscious of what is going on around them, or so absolutely lost to the outer world, that you may prick, pinch, or burn them, and they will not feel it, or know anything about it.

At St. Luke's Hospital I had a number of cataleptic cases under my care, and on one occasion as many as five were in the hospital at the same time. All the cases recovered with one exception, and this one died, and to my great regret the death took place whilst I was out of town; and though I most particularly requested that a *post-mortem* examination should be made, I found on my return that it had not been done. These patients generally exhibited in the most marked manner, the characteristic unconsciousness, and the muscular rigidity, and they usually remained, for a longer or shorter time, in the position in which they were sitting or lying, almost as rigid as bars of iron and as unconscious of the outer world as statues—in some instances so rigid were they that, like dutch dolls, they remained in whatever position you placed their limbs, and in other instances, they laughed, spoke, sung, walked, danced, or committed unseemly violence, at which they were astonished upon their return to consciousness.

I had the opportunity of observing a most interesting case of catalepsy lately. The patient was in the Infirmary for Epilepsy and Paralysis under the care of my colleague, Dr. Edward Meryon. The patient used to lie down and become unconscious of surrounding objects, and I many times placed her limbs in every possible variety of position, in which they always remained fixed as though they were set upon iron bearers. Whilst in this state I frequently spoke to the patient, but she did not hear; I pricked and pinched her, but she did not feel; yet she would frequently whilst in this condition, repeat conversations that passed between herself and other patients, and sometimes she would sing—often, over and over again, she would sing the morning or evening service of the church through from beginning to end:— but on recovering her consciousness, she was always profoundly ignorant of all that had occurred during her seizure.

A remarkable case was reported, I think by Macnish, whose work on sleep, though now a little old, yet contains the notes of many interesting facts. The case referred to was that of a girl who used to become unconscious of what was passing around her, and in this state would preach sermons. Her sermons, however, consisted of quotations and texts of scripture strung together, and after the phase of mind in which she performed these freaks had passed away, she remembered nothing of what had occurred during its continuance.

Sometimes the cataleptic will be unable to speak whilst the seizure lasts, but upon recovery will tell you of conversations which have taken place during the time of the attack. Such conversations, you will observe, have generally occurred during the recovery stage, and must not mislead you in your diagnosis.

In men catalepsy is very rare. In St. Luke's I saw one or two cases of imperfectly developed catalepsy in young men, but like hysteria in men they were somewhat anomalous. I believe one or two cases have been recorded, but the condition is uncommon.

You should always bear in mind that epileptics frequently have attacks at night, and when nobody sees them. Among my patients at the Hospital for Diseases of the Nervous System, I have several cases in which the attacks occur only at night;

indeed the fact of nocturnal seizures is of the utmost importance, because they are so often followed by mental aberration. The mental disturbance may last for a day or more, and it is by no means so uncommon that you should be surprised to find it.

An excellent illustration, and withal a very obscure case, was that of a boy in St. Luke's Hospital, who was the subject of night seizures. The attacks were followed by extreme melancholy, but the epileptic character of the attacks was not suspected by those under whose care he had been before his admission, in fact, his case was stated by a doctor who signed one of his certificates, to be melancholia, with frequent lucid intervals, and as this gentleman added to his certificate, "he has a lucid interval to day," the lad was ordered, by the Commissioners, to be forthwith discharged. He had, however, an attack of depression on the day of his discharge, and his friends sought, and obtained new certificates, and brought him back. I watched him very closely, and used to find, that every fortnight or three weeks he would wet his bed, complain in the morning of an intense headache, represented only by, and far worse than that of an over-night's debauch in wine, added to which he exhibited most perfectly the characteristic petechiæ, and observing these, together with the other symptoms, I assured myself of the epileptic nature of his malady. The melancholia which followed was extreme, and for three or four days his state was, in the utmost degree, that of abject misery.

In the intervals he was usually cheerful and happy, and exhibited a very fair share of intellectual power.

Epileptic cases, as we have seen, are sometimes very violent, and when affected as they often are with hallucinations of devils, they are termed demonomaniacs, and the wildness of the excitement, and the dangerous proclivities of these patients, are hardly to be described. For the time being, they more nearly resemble enraged wild beasts than human beings; one such case, a male patient, whom I saw some years ago, required almost constant seclusion, he was walking about a padded room absolutely naked when I saw him, and he was altogether unmanageable, if a less number than four men attempted to secure him, when it was necessary to enter the room for the purpose of removing foul bedding.

The stage of mental aberration sometimes precedes the fit as I before mentioned; more commonly it succeeds it, and may continue for a long time afterwards. Sometimes the patients remain more or less insane during the whole of the intermissions, and sometimes they speedily return to a normal condition of reason, and are able to follow their ordinary avocations, and conduct themselves as rationally as the most sane person, during the whole of their period of freedom from attack; but should such a case, suddenly commit some act of violence, his irresponsibility must be maintained, and although sad and melancholy acts result from the dictates of the hallucination, impulse, and perverted reason, induced by this deplorable form of malady, the law, in its justice, will certainly excuse, upon its being shown that the commission of the act was clearly beyond the patient's control.

LECTURE V.

Varieties of Mania. *(Continued.)*

Varieties of Mania so-called—Primary and Secondary conditions—Symptomatic Mania—Recurrent Mania—Accidents—Insolatio—Cardiac Disease—Puberty—Climacteric Period—Masturbation—Puerperal State—Menstruation—Metastatic Insanity—Phthisis—Syphilis—Acute Disease—Rheumatism and Fever—Gout—Habitual Discharges and Cutaneous Eruptions—Pathology of Chronic Mania—Treatment of Mania.

As I stated in our first lecture, the actual cause of insanity in all cases is, mal-nutrition of the brain, and it is therefore but reasonable to suppose that when a brain becomes badly nourished from any cause, insanity may appear as a symptom. In this, taken as a fact in the history of disease, there is nothing new. Many diseases appear as both primary and secondary conditions, a notable example being that of epilepsy. It is within the common experience of all of us, to find symptomatic and idiopathic epilepsy, and the same thing occurs in insanity. We must look therefore for insanity as a secondary or symptomatic condition, and among the so-called varieties of mania, we shall find numerous diseases, expressing themselves as mania; we find in fact many primary diseases of which mania is a symptom. The abnormal mental phenomena we recognize as mania being really the symptom of a particular state of brain, whether that state of brain be set up primarily as original disease, or whether it be secondary to some other condition, as for example blood poison.

Recurrent Mania.—Mania frequently recurs in attacks, which has given rise to an idea that there is a distinct class to be designated recurrent mania; but it is an error to consider recurrent mania as a class, the recurrence being dependent very much upon the exciting cause. You saw, the other day at Peckham, a very acute case of mania, which was associated with menorrhagia. The patient, you heard also, had already suffered from many similar attacks; that they always recurred at her menstrual periods, and with menorrhagia, and that when she was free from menorrhagia she was comparatively well, but this did not show the mania as a distinct class to be called recurrent.

The mania we have considered as complicated with epilepsy, is recurrent in so far, that it recurs in attacks which are recoverable, and so may any attack of madness be, which is dependent upon a condition of brain, in which the cells are not to any great degree structurally damaged, but which, as in the case of epilepsy, is under the influence of a recurring exciting cause. A weakened brain, if constantly disturbed by a recurrent disorder of some viscus, may exhibit mental aberration in recurring attacks, but there is nothing peculiar in the insanity or in its recurrence requiring its isolation into a distinct form or class.

Accidents.—It occasionally happens, that accident sets up changes in the brain upon which insanity follows, showing itself as a primary disease. But the insanity is really only a symptom of a particular condition of brain. The accidental circumstances most commonly recognized as likely to be followed by insanity, are, blows on the head, insolatio, cardiac disease, syphilis, and perhaps in some cases fever.

I mentioned incidentally in speaking of the etiology of acute mania, a case in which a predisposition had been set up by a sabre-cut on the head, and such cases together with those originating commonly from blows, are often grouped together. The group has some distinctive features, but the mania does not differ from mania generally. The cases do not as a rule come under treatment until they are chronic, and probably hopeless. But if you can learn the facts you will discover that in most cases the patient has during two or three or more years been subject to attacks and paroxysms of mania, very acute in form, but very temporary and fleeting, the excitant of the attacks being slight, such as small annoyances or mental strain, little anxieties, or even a glass of wine in excess of the wonted quantum.

Insolatio.—Sun-stroke is a cause of insanity, but a predisposing rather than an exciting cause; the insanity following sun-stroke is not necessarily mania. When mania is the form assumed, it is often very acute, sometimes it is very chronic, and sometimes even it is intermittent. It is more common to find sun-stroke a cause in patients who have been in India, than in those who have always lived in this country. Insolatio leaves behind it a damaged brain, and it only requires an excitant to develope madness in it, the form of which will vary according to circumstances,—the severity of the sun-stroke, the amount of change produced in the brain by it, and perhaps the length of time between the sun-stroke and the evolution of the insanity may all combine to modify the case, and produce acute mania in one case, chronic in another, dementia in a third, dementia paralytica in a fourth. The mania however when produced presents no special features.

Cardiac disease.—I have seen a few cases of mania in which valvular heart disease was present, and in which the morbus cordis preceded, and, as far as ascertainable, was the cause of the mania. As a rule, however, the mental disease associated with cardiac imperfection is melancholia or dementia. Two cases in point, were under my care in St. Luke's Hospital both of whom died, and afforded an opportunity of confirmation of the diagnosis by *post-mortem* demonstration. In each case the mania was very well marked by noisiness, violence, restlessness and wet and dirty habits. The mania in each case was very acute, both had suffered from acute rheumatism, and evidence of hereditary insanity from their histories was not attainable. It is not difficult to see how a diminution of blood supply to the brain, may result in wasting change, of which mania is a common symptom. The same cause, viz, diminished blood supply and wasting, probably is the origin of the insanity sometimes associated with chorea.

Puberty.—Insanity occurring at puberty usually shows itself in the form of mania, but ought not to be considered as a variety. Puberty is a period of physical susceptibility, maturation is rapidly progressing, and the child is developing into manhood or womanhood, at the expense of his or her stored-up resources. This is the time when any delicacy of constitution is likely to make itself known, and when any hereditary predisposition is likely to proclaim its presence. The weakness attendant upon puberty may therefore excite mania in a child predisposed to the malady. As a rule, the first attack of mania occurring at puberty is mild, but it has no specific difference by which it can be recognized as a variety.

Climacteric period.—The same cause operates at the climacteric period—another season when change is rapidly progressing, and when the attendant weakness is often sufficient to set up a maniacal state in a patient predisposed to it.

Masturbation. Mania associated with masturbation, is by some considered as a variety, but I am not inclined to consider it in this light. Masturbation is a very common practice with some lunatics, and is a cause of exhaustion which depraves the vital energy to the utmost, and is often the cause of an attack of mania becoming chronic or hopeless. Masturbation may be the excitant of an attack when there is a predisposition; but masturbation, *per se*, is not a cause of insanity. It is practised by the young of both sexes to an incredible extent, and particularly amongst the youth at boarding-schools, but insanity seldom comes of it. It rarely produces mania, and only does so in cases where predisposition exists;

and after a time both boys and girls who have practised it at school give it up. When it is commenced by a lunatic, it is a bad symptom, and you may expect that it will protract your case if it does not render it hopeless.

Again, the described varieties which pass by the names of satyriasis, nymphomania, hysterical mania, sympathetic mania, and sexual mania are not independent varieties: like the insanity following masturbation, they may follow upon irritation of the sexual organs, and the consequent exhaustion will set up the attack in predisposed subjects; the specific excitement is only a part of a pre-existing nervous disease.

The sexual organs, being associated with the essentially animal instinct of reproduction, are, more than any others, liable to excitement in the earliest stages of nervous degeneracy, and to induce a degree of mental degradation amounting in some cases to shameless indecent effrontery. I have seen symptoms of nymphomania in a girl of seven,—a young lady who took delight in openly and unblushingly seeking to gratify her desire. I have seen her look round a room, single out a stranger among the men present, go up to him, and immediately place herself in a position to excite her genital organs, but this and all allied conditions, are the outcome of an insanity already existing, and are not in any degree the cause of the insanity. Further, the sexual excitement is but a symptom, and contains nothing specific in it to constitute a variety. Sexual excitement is a common associate of insanity generally. In fact, most cases of insanity, develope sexual excitement at some time as they run through their course.

Puerperal state.—The puerperal state is a common excitant of insanity, when a predisposition exists. The predisposition is almost always hereditary, and the insanity is often mania, though dementia and melancholia, are often frequently seen. The mania is usually very acute and generally recoverable. Dr. Batty Tuke, of the Edinburgh Royal Asylum, favoured me some time ago with a pamphlet in which he detailed an enquiry he had made into the subject. Dr. Tuke classified his cases, according to the time of the appearance of the mental symptoms; of the 155 cases he examined, he describes 28 or 18·06 per cent., as insanity of pregnancy, 73 or 47·09 per cent., as puerperal insanity, and 54 or 38·8 per cent., as insanity of lactation.

The insanities associated with the puerperal state bear, it is true, some distinctive features: as the arrest of secretions and of fluxes ; *e.g.* those of the mammary glands, or the lochial flow.

These features are, however, no part of the insanity, and do not alter or determine its character.

They are physical symptoms which are to be regarded as evidence of a depraved vital condition and point to exhaustion as the cause; but they do not themselves induce the depraved vital condition, or modify the abnormal mental state.

Menstruation.—Disorders of menstruation are a common associate of mania, but are not in themselves causes of mania. Neither is the mania associated with disordered menstruation a class: like the insanity of the puerperal state that associated with abnormal menstruation follows in the predisposed upon a depraved bodily condition of which the disordered menstruation is only another evidence.

Metastatic Insanity.—This brings us to a division which has been called metastatic, though here again we are merely dealing with a word used in a classification based upon a supposed cause of the disorder, and not with a difference in the mental symptoms to be recognized by the adjectival application. The recognition of the various causes in the so-called metastatic insanity is, nevertheless, of the greatest value in treatment, and the division includes the cases which are seen to follow the suppression of accustomed discharges or the disappearance of eruptions. Many of the cases furnish good illustration of the symptomatic character of acute mania.

Mania sometimes appears during the course of an acute disease, as acute rheumatism, ague, gout, chorea, erysipelas, or any exanthem; sometimes it appears with chorea, and sometimes with phthisis.

Phthisis.—The insanity associated with phthisis pulmonalis is usually maniacal, and it is often very acute.

It has been regarded as a metastasis, as the pulmonary affection is often less active when the cerebral affection is at its height. This is a question of no importance, it is simply a phenomenon bearing out the pathological rule that if you do not find two inflammatory processes equally active in the same body at the same time, as you can divert the activity of an acute inflammation by a counter irritant, such as a blister; so, when an active change, associated with mania, occurs in the head of the subject of pulmonary phthisis, the severity of the thoracic symptoms abate. But there is another factor for consideration. Phthisis pulmonalis is an exhaustive disease, and exhausts the brain among other organs; and a brain in which a predisposition resides, when exhausted by Phthisis pulmonalis, will exhibit mania in the same manner as though exhausted from any other cause. I cannot give my assent to the opinion that an intimate relationship exists between insanity and phthisis. Schrœder Van der Kolk was of opinion that hereditary predisposition to phthisis might develope into insanity, and on the other hand, that insanity might predispose to phthisis. Dr. Clouston of the Carlisle Asylum, has also put forward the same idea, drawn from the statistics of the Edinburgh Royal Asylum. If phthisis predisposes to insanity why do not a large number of phthisics die insane? It is true that a large number of patients in asylums die of phthisis, but this is only part of the fact, that phthisis kills a very large percentage of the population of the country. The two pathologies may exist in the same individual; a person may be born with a predisposition to both insanity and tuberculosis, may develope phthisis pulmonalis, and on becoming exhausted may become insane; but to suppose that there is any other connection or relation between the two pathologies, I believe to be altogether erroneous.

A number of statements have been recorded as to the fact, that 25 per cent. of the deaths in asylums are from phthisis. Dr. Clouston calculates the percentage as 60.* We want further statistics on this point. Dr. Boyd, of the Somerset Asylum, found only 61 deaths from phthisis in 302 *post-mortems* of the insane, or about 20 per cent., which is nearer the proportion of deaths by phthisis in the sane, and I think will be found to be nearer the fact.

Syphilis.—Insanity is sometimes associated with syphilis, and may arise primarily from syphilitic change in the brain, or it may happen that primary syphilis becomes the excitant of the attack in a person already predisposed; in the latter case a moral exciting cause very probably is added to the physical and exhausting.

Acute disease.—The mania attendant upon acute disease is generally very transient, arising from blood-poisoning, and disappearing when the fury of the disease has expended itself.

Acute disease, as rheumatism or fever may, however, leave the brain exhausted, and insanity which is sometimes, under such circumstances, persistent, may follow, though such insanity is usually melancholia.

It has been advanced by some pathologists that

* Dr. Clouston's paper. *Journal of Mental Science*, April, 1863.

mental symptoms occur only in those cases of acute rheumatism in which the patients have been drunkards, but my own observations are opposed to this view, though my cases are not numerous. One was in a lady to whom the suspicion of drunkenness could not attach, and two were in children who never had the opportunity of obtaining strong drink. That predisposition to insanity existed in all their cases I have little doubt. In the case of the lady it was clearly hereditary, and I would join issue in the observations of the pathologists who have described wasted brains in such cases; but the brain may be wasted from causes other than drink, and if alcohol is a frequent cause of cerebral complication in acute rheumatism, it certainly is not the only cause.

Gout, like rheumatism and fever, may set a light to a mine of insanity. The brain will become imperfectly nourished by reason of the gouty change in the blood, but my own experience leads me to the belief that gout is only an excitant of insanity in those already predisposed, its attacks are intermittent, and sometimes alternate with attacks in the great toe and other parts.

You saw an illustration of this fact in a patient at Peckham the other day; and I have had a gentleman under my care for some years who has alternate attacks of mania, gout in great toe, and bronchitis, all of which disappear under the use of colchicum. He has, in his family history, very direct evidence of the fact that he has inherited both gout and insanity.

The attack of mania associated with gout, is, as a rule, more or less easily and rapidly recovered from as the exciting cause disappears.

Ague is another of the diseases which sometimes shows itself as mania, but you can, by the patient's history, usually discover the existence of the ague, and you must, of course, treat the case accordingly.

Alternate attacks of mania and skin eruption are not by any means uncommon. Some of you saw a case here in the hospital a short time since, in which symptoms of mania followed upon erysipelas, and it is in alternation with erysipelas that the occurrence perhaps is most common.

I have seen one case in which the attacks alternated with very typical psoriasis, and in this case there was very clear hereditary history. There are numerous similar instances on record, as also the record of cases showing that the arrest of some habitual discharge, as a seton, or a discharging scrofulous gland, or of the drying up of an exzema, has been immediately followed by insanity. These facts, however, cannot be taken as indicative of more than we have already seen, viz., the patient has a predisposition to insanity, and a very slight cause is sufficient to set up an attack.

Some writers have regarded the insanity associated with alcoholism as a class. We shall reserve the full discussion of this subject for a future lecture. Insanity associated with drinking assumes a variety of forms, none of them being peculiar to alcoholic excitants, unless it be the delirium of delirium tremens, and this is not constant. Alcohol will excite insanity in the predisposed, but that it will *per se* induce insanity in a healthy individual, has, I think, never been shown.

Pathology of Chronic Mania.

In our consideration of the pathology of the acute form of mania we noted that we were dealing with a dynamical condition only. But as regards chronic mania, the condition is not altogether dynamical, it may be statical; nevertheless, during life you have no positive means of determining whether the imperfection of the nutrition of the brain has been sufficient to induce tissue change, and thereby permanent damage, or whether the imperfection is only one of atony. It is, however, only in cases where the defect has never exceeded atony, that the apparently perfect recovery, after a number of years of mania, can take place; for if the brain is structurally changed, the cells, which during the first stage of the disease lost their faculty for comparison and correction, cannot possibly return to their ordinary and healthy activity.

The conditions of brain upon which chronic mania depends are not those of a fatal disorder, and the subjects of chronic mania usually die of some intercurrent disease. They may die of some cerebral disease, as cerebral hæmorrhage, unconnected with the origin of their mania; or they may die of any visceral disease, as kidney, liver, heart or lung change, from which they may have suffered independently of their mental disease.

In almost all, if not in all, subjects of chronic mania, you will find some foot-prints of the malady displayed, upon a *post-mortem* examination of the patient's brain. These foot-prints are often manifested in formative change, and give evidence that a low sub-acute form of inflammation has existed, by which the membranes have become thickened, and the areolar tissue, which, with the vessels, tacks the pia-mater to the brain, is increased. The thickening of the dura mater is sometimes enormous. I have collected specimens the thickness of ordinary chamois leather. Sometimes the skull is enormously thickened, as you may see from the specimens on the table: Nos. 1073[75] (museum numbers). The diploë has entirely disappeared, and the whole thickness of the bone is a dense mass of sclerotic bone-tissue. The cases were from the lunatic wards which formerly were attached to this hospital. Occasionally you see bony tissue developed in the membranes, and very commonly you find morbid adhesion of the dura mater to the skull.

I have already mentioned pachymeningitis as described by Schrœder Van der Kolk, but whether acute inflammation of the dura mater ever exists in association with mania is still open to question. Acute inflammation of this membrane is an exceedingly rare pathology, though several instances are detailed in the posthumous work of Professor Schrœder Van der Kolk, to which I have already alluded. The adhesion, the bony development, and the thickening may be regarded as evidence of chronic inflammatory change, but whether this chronic inflammation of membranes has its origin prior to the mania, or whether it is consequent upon it, may fairly be open to question. I am inclined to believe that it is a consequence, for we do not find any evidence of it in cases of death from acute mania. The sequence of the changes appears to be:

1st. Hyperæmia coincident with acute attack.

2nd. Impaired nutrition and tissue change in the cells, resulting in

3rd. A moderated but continuous hyperæmia coincident with chronic excitement, producing

4th. Formative changes, thickened membranes, bony deposits, and increase of areolar tissue.

Occasionally the two layers of arachnoid are found to be adherent, but this is comparatively rare, and arises from arachnitis, which may have occurred independently of the mania or its origin. In some cases the change begins in the skull and membranes, and precedes the mania which is attendant upon secondary change in the brain, and as striking examples of this class I may instance traumatic cases in which a predisposition is set up by changes which are consequent upon a blow on the head.

Under all circumstances the change is progressive; the thickened membranes impede the blood supply, press upon, and destroy the vessels which often become fatty and are sometimes absorbed; impairment of nutrition then necessarily results, or is increased when it already exists as a consequence of hyperæmia and paralysis of the cerebral vessels, and atrophy or wasting follows. With this wasting both grey and white matter shrink, and the convolutions sometimes become thinned down to the utmost degree. The shrinking of the convolutions seems almost like an attempt at retrogression to the primitive form, in which the brain, as yet untutored in intellectual operations, has not fully developed its convolutions.

The wasting of the brain has been remarkably well described by my friend and colleague (at the Hospital for Diseases of the Nervous System), Dr. Lockhart Clarke. He pointed out holes and tubules in sections of the white substance of the convolutions and of the optic thalami, which he regards as widened perivascular spaces. Some German writers have laid much stress upon this condition, and have described such brains as having a worm-eaten appearance, or as resembling Gruyere cheese; the condition is, however, only wasting of the brain and widening of the perivascular canals, and Dr. Clarke* adduces as evidence that these holes are widened perivascular canals, the fact that the walls of the cavities are perforated by the smaller channels which convey branch vessels, and that some of them exhibit, on examination, the sheaths of the vessels which remain behind, although the vessels themselves may have become destroyed or absorbed. Dr. Dickinson of St. George's Hospital has found the same change in the brains of diabetic pa-

* *Journal of Mental Science.* Jan., 1870.

tients. Drs. Batty Tuke and Rutherford,* have made some observations on the subject, and they seem to regard some of the holes as unassociated with blood-vessels, but due to solution of continuity of nerve tissue. My own observations lead me to endorse Dr. Clarke's statement. But I have placed some preparations on the table, so that you may have an opportunity of judging of the facts for yourselves.

Some of the microscopical conditions of brain, are exceedingly instructive, as they show, not only a diminution of the gray substance, but a disappearance of the caudate cells. Dr. Lockhart Clarke has described the cortical substance of the brain, as consisting of seven concentric layers of cells. You will find his paper on the subject in vol. xii of the proceedings of the Royal Society, 1863; and you will also find a clear and concise description written by Dr. Clarke, at page 60, of Dr. Maudsley's *Physiology and Pathology of Mind*, 2nd Edit. 1868.

That the cells change their character is evident. They lose their sharp outlines and become granular; and then either break up and appear as little heaps of granules, or else their contents are absorbed, leaving as it were a shrunken cell wall with attenuated prolongations.

Under the microscope are some sections of healthy brain, and also some of extreme wasting in which you will observe a very marked difference.

* *Edinburgh Medical Journal.* Oct., 1869.

Schrœder Van der Kolk asserted that he found in mania, that the cortical layer under the frontal bones was darker coloured, more firmly connected with the pia mater, and softer than in health.

You will often find a considerable quantity of fluid in the ventricles, and also an excessive quantity of fluid, on the surface of the brain; much importance cannot however be attached to this fact, for the fluid is effused as the brain shrinks or becomes atrophied. A more interesting condition is that of granulation of the ependyma, and particularly of the surface of the septum lucidum and 4th ventricle, a condition which was, I believe, first pointed out by Dr. Lockhart Clarke, who described it in Beale's *Archives* (No. ix, 1861), as hypertrophy of the epithelium.

It is an appearance which I have observed over and over again in the *post-mortem* room of this Hospital, and which I believe may be caused by any inflammatory condition, and may be set up by any acute inflammatory disease, as fever or pneumonia.

Sometimes the remains of old capillary hæmorrhages and cicatrices are found, and their presence is dependent upon degenerative changes in the vessels themselves. The idea which has been suggested, that they result from the stress of the congestion and the violence of the excitement, is, I think doubtful. However great the congestion may be, its force must be more or less broken by the rete mirabile, by which the retardation and equable flow of blood through the brain is secured.

When epilepsy has been present I have sometimes found conditions which were special, and directly associated with the epilepsy.

Simple wasting of the brain you should remember is a common cause of epilepsy, and in the essentially "epileptic patients" of our asylums, atrophy is always the condition of brain found after death.

In the class of cases, however, which I have described as containing demonomaniacs, coarse conditions of brain are often found, of which the most common are tumours pressing upon and destroying portions of the brain's surface. These may have their origin in the membranes; sometimes they arise in the brain substance itself. On a few occasions I have seen bony tubercules growing from the tubula vitrea, and encroaching upon, and destroying the brain substance. This preparation, No. 1072[61], was a case which I examined in this hospital: you see the tubercle is about the size of a marble, and it is projecting from the frontal bone; it was destitute of dura mater, but the membrane from the portions of the skull around are mounted, so that you see how the tubercle projected through. The patient was epileptic and maniacal. This drawing was made from the calvarium of a case I examined in St. Luke's hospital; the preparation I am sorry to say has been mislaid. As represented in the drawing, cauliflower-like tubercles grew inwards from the frontal bone, and the falx major was converted into a dense mass of bone. This patient was epileptic for a considerable time, then became maniacal, passed ra-

pidly into a state of dementia, and died of acute tuberculosis.

In addition to the changes already noticed, is one which was pointed out by Schrœder Van der Kolk, viz., a dilatation of the vessels of the medulla oblongata; this dilatation is confined to the venous vessels, and I think is well illustrated by the specimen I have placed on the table, and which you can examine for yourselves.

Although I have demonstrated very positively some of the coarse conditions of brain and skull, associated with epilepsy, you must not forget that epilepsy is a dynamical condition, and may be set up by a variety of causes, coarse disease not being at all essential to its production.

Great as are the advances we have lately made in our knowledge of brain physiology and brain pathology, our science is still in embryo, and we have yet to learn much of healthy conditions, particularly of the physiology of the cells, before we can isolate any set of these, and say, from definite appearances, " these are the cells of madness." The changes we have recognized we have not demonstrated to be constant, for we have found some of them from time to time in subjects whose history has not revealed any mental disease. We have much to learn, and much to unlearn and correct, in regard to nerve pathology. But one fact we may state with absolute certainty, viz., that the change is one of perverted nutrition. This usually arises in abnormal vascularity, which tends to structural and tissue changes, and general retrograde metamorphosis.

Treatment of Mania.

The first question we have to decide is, whether the case can be treated at home or not.

The general answer to this question is, that a patient may be treated at home if facilities are within reach. There are some cases in which home treatment may be pursued with much advantage to the patient; but there are very many circumstances which render removal, in a large number of cases advisable, and in some it is absolutely necessary.

Having regard to many common exciting causes, as those associated with local or domestic disaffection, it is evident that the first indication is removal of, or removal from those causes. We must next attempt to arrest exhaustion, then endeavour to allay the irritability of the brain, and promote its nutrition. The therapeutic agent most valuable for the accomplishment of all these indications is physiological rest, and there is not the least doubt, that the best mode of obtaining rest is by keeping the patient in bed.

You see a recent acute case for the first time, and you find the patient morbidly active, walking constantly about the room, perhaps sitting down first on one chair then upon another, and so on until he has tried all in the room, or he walks about gesticulating, muttering, talking, screaming or shouting, and on enquiry either from the patient or those about him, you learn that his restlessness continues day and night, and that at times he may even be more restless than at the moment when you see him, you may learn too that for many nights, perhaps weeks before the commencement of the attack, the patient had been sleepless. Perhaps he has had some anxiety over which he has brooded, abstaining from food, and disregarding the calls of nature. Such a patient should at once be ordered a warm bath, and should be put to bed, and you will often be astonished at the benefit this treatment will in a few hours produce. The soothing effect of the warm bath in acute mania is very remarkable, and you will often see that your patient, after a warm bath, will become calm and go to sleep.

The *modus operandi* of the warm bath is physical, and undoubtedly its first action is to relieve the capillary congestion of the brain, by determining the blood to the skin, and then by promoting the secretion of the skin glands to inaugurate the commencement, or at all events the attempt at the commencement, of a more normal state of the glandular organs generally.

In the place of the warm bath, you may sometimes pack your patient in a wet sheet with infinite advantage. The wet pack calms excitement very much in the same manner as the warm bath and bed. The element of shock produced by contact of the skin with the cold sheet, is of doubtful value, but the sheet soon becomes warm, or a sheet wrung out with warm water may be used. The wet pack induces an action of the skin of more violent character than the warm bath. The wet pack, too, furnishes another element which is of use in mania, it confines the limbs and so enforces physical rest which otherwise the general irritability would prevent.

It is, however, necessary to warn you, that the wet pack is dangerous, especially if the patient labours under a damaged heart or defective circulation. The amount of sweating resulting from the wet pack, too, induces intense thirst and the patient will, if allowed, drink a fatal draught of cold water. You must take very especial note of this or you may find your patient suddenly collapse, if you yield to his beseeching. The cause of death in such cases appears to be, and probably is, from shock to the semi-lunar ganglia of the sympathetic. After a wet pack the patient may have a shower bath or a cold sponge bath, or he may have a cold or nearly cold plunge bath, if you can keep him from drinking the water; he should then be rapidly dried and put to bed.

Dr. Lockhart Robertson, one of the Lord Chancellor's visitors, was, when Superintendent of the Sussex Asylum, the great advocate of the wet pack, and he kindly gave me the opportunity of witnessing his method on one occasion when I was visiting him at Hayward's Heath.

Dr. Robertson's method:—a sheet was dipped in a pail of cold water, and wrung; it was then laid upon a mackintosh sheet which covered a mattress, the patient was then stripped and laid full length on the sheet with his arms drawn straight beside him, the sheet was then rapidly folded over him, and he was then covered with a blanket, which was also wrapped tightly round him.

Dr. Robertson recommended that after a patient has been in the sheet for an hour or an hour and a half, he should be unpacked, have a couple of pailfuls of water thrown over him, be then rubbed down with a damp sheet and again packed, and that this treatment be continued for several hours, or repeated at intervals during the day.

You must remember that the remedy is a violent one, and that it requires much judgment in its administration, and finally that the duration of the treatment must not be regulated by rule but by the condition of the patient at the time of application.

For the sake of conciseness I may assume that you are called upon to treat a patient attacked with acute mania in his own private house. How are you to manage the case? In the first place, you must direct that the patient shall not be left for a single moment without an attendant on account of the facilities afforded in a private house for doing mischief. But even in an asylum, where you have architectural arrangements for preserving a patient from the injury which his insanity might lead him to attempt, it is well that he should never be left. The reason for this is obvious, when you remember that the influence of a sane mind upon a lunatic is often immense. Generally, a patient gains confidence and some power of self-control when he feels himself under the influence of a stronger mind than his own, and should he waken from sleep and find himself still under the governance of some person upon whom he begins to rely for support, he will not attempt to satisfy his morbid impulse.

The surroundings of a bed-room may become the basis of illusions, and, if a patient who is the subject of an imperfectly acting brain should wake from sleep and find himself alone, the illusions formed at the waking moment may become the origin of fresh outbursts of excitement.

As to the surroundings, they should be as simple as possible, and the less furniture there is in the room the better. It is often advisable, in a very acute attack, to make up a bed on the floor, and to remove from the room bedsteads, chairs, tables, washstands, and all articles with which the patient might hurt himself or do damage. He may with advantage have two rooms set apart for his use, the one dismantled in which he may be kept during the period of violent excitement, the other furnished and comfortable to which he may be removed as soon as he is calm. In his restless and excited stage he should have some light and loose but warm clothing, as he will not, in all probability, allow himself to be long covered by the bed-clothes. It will, in some instances, be necessary to make this clothing of very strong material, such as canvas or ticken, as the patient often feels all clothing irritable, and will tear any ordinary garments to pieces in order to get them off. It is, however, necessary to preserve the patient from cold, and it is to his positive benefit to keep the surface of his body warm; therefore the strong clothing should be wadded and quilted, and secured with leather straps and locked buckles.

A material known as swan's-down cotton is an excellent lining for a ticken dress.

If you can select the room, choose one on the ground floor, with shutters to the windows, so that you may exclude the light if necessary; you should be careful, too, that the window is fastened, so that the patient may have no facility for escape.

Having arranged for the patient's safety, and provided a good and reliable nurse, you should direct that advantage be taken of every calmer moment to try and court sleep. The patient may often be induced to lie down, and if only for a few minutes at a time it is a gain; further, he may be soothed with tender nursing: he may be comforted with cold cloths applied to the temples or forehead, or by allowing him to rest his head on the nurse's lap, and with gentle persuasion he may be induced to lie still. He should not be visited by any person for whom he has assumed a dislike or aversion, and as a rule he does best if he is not visited by any of his relatives during the very acute stage.

It very frequently happens that the patient becomes more or less disaffected towards his relatives, and their presence irritates and excites him; relatives, too, often cannot assume the same control as strangers. You must, however, use discretion in this matter, for sometimes the patient will have confidence in and obey a relative in whose care he may with safety be left.

Custom and usage have prescribed male attendants for male patients and female nurses for female patients. You will find, however, that a good female nurse will sometimes gain the confidence of a male patient and soothe him better than a man can, and I invariably prefer a female nurse for an acute case unless the patient be so violent that the assistance of a man is indispensable.

The nurse must, however, be intelligent. A stupid old woman will do more harm than good. A nurse may display her tact in the manner in which she induces the patient to take food, and food in enormous quantities is required by maniacal patients, but they will refuse it, and declare that it is poisoned. By timely persuasion and embracing of opportunities, considerable supplies of nourishment may be consumed by the patient. The nurse must, however, study to hit off the right time.

You will often find that a female patient will

take food from your hands, when she will reject it if offered by the nurse, and when you find that the patient has confidence in you rather than in those about her, you will be likely to succeed best in her case by encouraging this confidence, and feeding her or administering her physic with your own hand, until the acute attack has passed. So essential is it to feed the patient, that every artifice to induce him or her to take food is justifiable.

In mania, persuasion is generally sufficient to induce the patient to eat, and mechanical feeding is not often necessary, as in melancholia.

When you have provided for the patient's general comfort, the great necessity is to procure sleep, and you will be sorely tempted to give opium, or digitalis or Indian hemp, or some other neurotic, but I cannot in too strong terms deprecate the use of sedatives in mania.

Opium checks the secretions which it is your object to promote, and in acute mania it rarely produces sleep. Moreover, if it does so, it accomplishes the end only by carbonising the blood, it does not unload the congested vessels of the brain, or release their paralysis. All it does is harm, for it deteriorates still further the atonic and unnourished nerve tissue. So strongly have I felt upon the subject of opium, and so convinced was I of its harmful effect upon the maniac's brain, that I never once prescribed the drug in any form during the whole time that I had charge of St. Luke's Hospital, and I am convinced that my patients were the gainers by its avoidance. The vital power of a healthy person under the influence of opium is depressed, and the effort required to throw off the poison, further depraves the nervous system. How much greater then must that depravity be, when the nervous energy is previously impaired by disease?

Next to opium I would mention digitalis, a drug which has been much vaunted in the treatment of mania, but in regard to which I would urge objections as strongly as when speaking of opium. I have given digitalis in doses varying from 10 minims to half an ounce, and I cannot say that I have ever seen any good result from its use. Sometimes the excitement has been a little moderated with the depression of the circulation, but its violence has always returned when the effect of the drug upon the heart has passed away, and it is evident that you cannot improve and nourish a brain already depraved and depressed, by poisoning the blood even though you temporarily lessen the tendency to congestion.

The action too of digitalis as a drug is very uncertain, and it may produce syncope and even fatal syncope when you least expect it. I was once summoned to see a patient for whom I had prescribed digitalis, and who had taken the drug and borne it well for many days. On reaching him I found him in an alarming state of syncope; fortunately the symptom soon passed away and the pulse returned to its normal state: but I may mention, that notwithstanding the fact of this great depression, the syncope had no sooner passed away than the excitement re-appeared with renewed force and violence.

Henbane is a drug to which the same argument applies, but its action is more uncertain than digitalis. I once gave to a sane patient a dose of an ounce and a half of the tincture without producing the slightest effect upon him, and as a sedative for the insane I think it worse than useless.

Cannabis Indica as a sedative in mania generally, is a drug as much to be condemned as any of the foregoing, but in mania associated with epilepsy it is sometimes of use, and it is also valuable in mania when associated with menorrhagia.

Tartarised antimony has been much praised, but I have never seen any real benefit from its use, though I have frequently witnessed its internal administration and also its application to the skin. Hydrocyanic acid too, was at one time much used, but I have very little experience of it. I am inclined to regard it as a more or less harmless sedative, and it may possibly be of service; nevertheless if you use it you must watch your patient very carefully, for it is a very uncertain drug.

It was strongly advocated by Schrœder Van der Kolk, who also recommended bleeding in very acute mania. I should be inclined to go with Schrœder Van der Kolk if you could bleed topically; if you can bleed from, or apply a leech to, the nose, you may relieve the plethora of the head, but general bleeding is to be condemned in strong terms, as it weakens the patient and necessarily aggravates his malady.

If from any cause a sedative in mania is necessary chloral is undoubtedly the best. If given at all it should be given in full dose, and the best vehicle for administering the drug is beer. There are some cases in which the sleep induced by chloral or chloroform becomes natural when the effects of the drug have passed away. It is, however, the rest and not the chloral that restores, though the drug often gets credit for more than its due.

I have given chloral in very large doses without producing sleep, and though in some instances I have found it of value, in others it has appeared to me to be absolutely useless. Of all medicines, however, chloral is perhaps the one most to be relied on.

A drug of real value in the treatment of mania, is to be found in alcohol. I have rarely failed to find it give balance to the most tottery brain. It will calm the most restless and excited mania, and will almost always induce sleep, even in very obstinate cases. It matters little in what form it is administered; wine, either port or sherry, may be given, or brandy or any other spirit is equally useful, and may be administered without stint, in any manner in which the patient can be induced most readily to take it. My practice in St. Luke's Hospital was always to order wine or brandy, with beef-tea or eggs, for a restless case, and this treatment was eminently successful.

There is reason to suppose, as I have already told you, that in mania the brain is more or less congested, and its vessels in a greater or less degree paralysed, and the evidence furnished by the results of treating cases upon this theory favours the supposition.

Alcohol in excess will congest the brain, but within a certain limit will release paralysis of the cerebral capillaries, and unload them when congested.

I have no doubt most of you have had some experience of the temporary paralysis of the capillaries of the brain, caused by over-study, or over-reading. You may have been sitting up late at night, whilst reading for an examination, and on retiring to your bed you have found yourself restless, excited, and sleepless, with a sort of conscious cerebration, in which the subjects of your study have been constantly and painfully revolving in your mind; this condition arises from a congestion and temporary paralysis of the cerebral vessels, which a glass or two of wine will speedily relieve.

You will probably hear very many objections raised against alcohol. The patient's friends will object to its use, upon the idea that it will excite the patient still more, or that it will feed the patient's craving for stimulants, or perhaps the patient will object and declare that it causes headache; nevertheless persevere, the headache and the excitement are the first fruits of weakness, which the alcohol will enable you to combat, and if the first dose does not calm the patient, give more. It is better to give it with food when practicable, and if the patient refuses solid food you should give brandy, or wine, mixed with milk or eggs, or arrowroot or gruel; failing these try beef tea or jelly, but remember, that neither beef tea, nor jelly contain much nutriment. If you succeed in getting the patient to take some beef tea and wine at one time, you may be able to induce him to take a mixture of beef tea and minced meat, and either some wine or some brandy with it, at another.

There is another therapeutic agent of value in quelling excitement, viz., darkness; a darkened room will often soothe the greatest amount of violent excitement. As a rule I have found maniacal patients much comforted by removal from the influence of light. Light, often is painful to a patient's eyes, and by paining and irritating them continues or feeds the excitement, but when removed to a darkened room, or when the rays of light admitted into the room are moderated a stimulus is removed, and sleep, which above all things is to be courted, is often thereby induced. You must be guided by circumstances, whether the light be shut out with shutters, or moderated by thick curtains; there is no necessity to render the room absolutely dark, and of course, if a patient protests that he or she is frightened when in a darkened room you must not push the treatment. I have, however, very rarely found any objection to darkened rooms, and in St. Luke's Hospital I invariably employed this method of treatment whenever the excitement was at all persistent or obstinate.

You will often find darkness of great value in

those cases of chronic mania, in which occasional paroxysms of violent excitement occur and last for a time, varying from a few hours to a few days. Such patients derive the utmost comfort from being placed in a darkened room, and you may depend upon the sedative influence of abstraction of light as the readiest means of cutting short such paroxysmal attacks.

In acute mania, recovery is usually marked by a return of sleep. The patient passes into a sound sleep and wakes calm. His ideas may be confused, and his delusions may be prominent, still the excitement has abated, and the initial stage of convalescence has been entered upon. As soon as the acute excitement has passed, the patient may be removed to a furnished room; but he should be kept in bed and treated as an invalid. He must be carefully watched, well supplied with food, and he must have stimulant enough to support his strength, but the quantity may be diminished as he improves. A daily bath will be essential, it may be administered either at night or in the morning. It may be given in the morning if the patient sleeps well, otherwise it should be used at night.

As his general health returns the patient may get up, and after a little while he may be taken out for a drive or walk; and you will often find that gentle open air exercise is exceedingly beneficial. You must watch the bowels, and if necessary give an occasional purge, such as grain or two of calomel with colocynth and hyoscyamus. If you are fond of tonic treatment you may give a little mineral acid: small doses of nitro-hydrochloric or of phosphoric acid may often be used with much benefit.

Chronic mania, you will often find, is very difficult to treat, and the difficulty is obvious from the pathology of the disease. The brain tissue is undergoing change. The condition is active, and the excitement although low, is continuous. The patients are usually more excited at night than in the day time, and they will often occupy many hours in talking, shouting or singing. I had a patient under my care who rarely spoke in the day time, but who used to talk the whole night through. His sleep could only have been obtained by snatches, yet his general health did not suffer.

Some chronic patients are wet and dirty; sometimes this filthy habit arises from a morbid delight in giving trouble, sometimes from forgetfulness and sometimes from the severity of the nightly excitement. Wet and dirty habits in chronic maniacs can usually be overcome by a watchful attendant, or in the mischievous cases by little rewards or privations. A patient will, often, have a weakness for tobacco or snuff or jam or fruit, and you may indulge him in these luxuries when he is clean, and deny them when he is wet and dirty. He will soon improve. If his periodical excitement and restlessness become great, you may help him to go to sleep with small doses of chloral, or a little alcohol. Numbers of people in good health take some stimulant before going to bed, and should they become insane they have all the more need of their night cap or help to sleep.

You will sometimes find that semi-chronic cases are very destructive; they will pick the bed clothes to pieces, and defy almost all attempts to stop them. Their destructive habits often arise from fidgetiness and a desire of occupation for the hands, sometimes the destructive habit is indulged in from mischief. The patient cannot sleep, and employs himself in destroying the bed clothes from the moment it becomes light. Such patients should occupy a dark room if possible, and their hands may be confined in gloves or in a pinafore, or an attendant may sit beside the patient's bed for a few nights until the habit is broken, or you may resort to the system of rewards and denials of luxuries, as I before suggested; the latter remedy will often be successful when everything else has failed.

Chronic maniacs should be frequently bathed, well fed, allowed plenty of open air exercise, kept free from excitement, and placed under good hygienic conditions, they will then be under the most favourable conditions for recovery.

It is of infinite importance that the bowels of all maniacs should be regulated, and you will find that they are often as obstinate as the patients to whom they belong. Sometimes a mild purge as castor oil will be enough, or an enema may be very useful. I have often found a turpentine enema give the patient great comfort, but as it only clears out the large bowel it is often insufficient, and sometimes a very powerful purgative, as croton oil or elaterium is required. A mercurial purge, however, often stimulates all the secretory and excretory organs beneficially, and its use may be supplemented by a constant saline purges; sometimes aloin and jalapin are very useful. So essential however is attention to the action of the bowels, that not only may the irritation of the brain be kept up by a loaded colon, but mania may actually be excited by it.

A boy of eleven years of age, was admitted into St. Luke's Hospital with symptoms of mania; he had frequented a Sunday-school, and on his admission into St. Luke's, constantly gave utterance to fragments of scripture usually involving the expression of the words, "Lord Jesus Christ." The child got worse instead of better and the case seemed to me to be very obscure. I made particular enquiries as to the state of his bowels, and both he and his attendant assured me that they were moved every day. Early one morning, however, the attendant came to me and told me that the boy complained that his bowels were confined, and I ordered him a soap enema; but the attendant shortly returned and said that he could not give the enema, and I went to the patient with the intention of administering it myself. To my surprise I found the anus widely distended and blocked with a large mass of dark brown dry fæces, which prevented the syringe from passing; I cleared away some of the fæces with the handle of a scalpel, and after removing sufficient to ensure closing of the *sphincter ani*, administered a soap and water injection, which brought away the astounding quantity of two chamber-pots full of hard scybala; it was evident that the boy's bowels had not been moved for weeks, perhaps for months. He got rapidly well and was discharged recovered.

When mania can be traced to any specific cause, as gout, rheumatism, or syphilis, the treatment must be directed towards these specific conditions and in such treatment, drugs will often be of value. Acute mania in a gouty subject, may disappear after a few doses of colchicum. And syphilitic patients will often improve under iodide of potassium or bichloride of mercury. Mania associated with acute rheumatism is an unfavourable symptom. Such patients must be purged fully and freely, you may use mercurial purges and salines with some prospect of success, but, in my experience, the prognosis of mania in the course of acute rheumatism is unfavourable.

In mania associated with epilepsy, bromide and iodide of potassium may be of great value, and in all cases of mania, where absorbents are indicated, you may find bromide and iodide of potassium useful. I always prefer using bromide with the addition of iodide in the proportion of about one grain of the latter to ten of the former, and in my epileptic practice I have found these drugs of real benefit, I have not, however, found bromide of potassium to possess that charm which some writers have attributed to it, nor have I found it so generally serviceable in either epilepsy or mania as many have made it out to be. I also find bromide of ammonium, and chloride of ammonium of service, and sometimes bichloride of mercury has proved highly efficacious. In cases of epilepsy in which absorbents are not indicated, or in which the origin of the disorder appears to date from a former generation, or the malady confirmed, and particularly when uterine disturbance is an accompaniment, I have found belladonna very useful in keeping the attacks at bay, and I usually give it in doses varying from $\frac{1}{4}$ gr. to 1 gr. of the extract mixed with $\frac{1}{4}$ gr. to 1 gr. of the powdered leaf every 6 or 8 hours. The *modus operandi* of belladonna in epilepsy is probably a lessening of the liability of the surface vessels of the brain to contract, because the drug tends to keep the venous vessels of the medulla oblongata constantly full. Such at all events is the theory, of its operation which Schrœder Van der Kolk propounded.

Nitrite of amyl is a drug I have used in epilepsy with eminent success, and I believe that further observations will prove it to be a very valuable medicine.

I need not now occupy you with any lengthened remarks upon masturbation. The habit when confirmed is a result of mental disease which it tends to aggravate, and therefore should be checked if possible. But the difficulty lies in the fact, that the patients are insane, and not amenable to reason. Appliances without end have been invented, and used, but all to no purpose, and even actual restraint is often useless, the patients will tell you that they can satisfy their craving by friction with the bed clothes even though you may confine their wrists and ankles. All attempts therefore to treat determined masturbators are I believe hopeless. The worst cases, both male and female, will shamelessly expose themselves and masturbate openly, and you will find much difficulty in preserving common decency with them. Such patients require and will consume enormous quantities of food, and it is best to try and restore their general condition with food in the hope, that with an improvement in their mental and bodily state the pernicious habit will be given up.

We must now give our attention to the subject of Melancholia.

LECTURE VI.

MELANCHOLIA.

Melancholia Described—Derivation of the Term—Definition of Melancholia—Forms of Melancholia—Cases of Melancholia—Post-mortem Appearances—General Condition of Melancholia—Sleeplessness—Exaggeration of Impressions—Hallucination in Melancholia—Morbid Religious Impressions—Sudden Impulse—Masking of Feelings—Suicidal Tendencies—Bodily Condition—Varieties of Melancholia—Pathology of Melancholia.

Melancholia has been discribed by many systematic writers under the head of monomania or intellectual monomania, but I have already told you that no such disease as monomania exists. Melancholia is now generally recognized as a more or less distinct form of mental disease, and it comes very properly within our natural classification.

The term melancholy derived from μέλας black, χολή the bile, is of very ancient date, and was in use long before the time of Hippocrates. The immortal work of Burton, perhaps one of the most learned works in the English language, has melancholy for its subject, and its celebrated author assumed the name of Democritus, a melancholic who was described by Hippocrates.

The word had its origin in the belief that the liver was the seat of the soul, and of intellectual operations, and it was supposed that the particular condition we call melancholy was consequent upon a secretion of black bile. Of course I need not tell you that the term is altogether valueless as defining the malady, and we only retain it since it is a word so generally comprehended. It tends, nevertheless, to show how accurate the observations of the ancients were, for although a secretion of dark bile is not the cause of the mental state known as melancholia, yet almost all forms of affection of the liver are accompanied with more or less mental depression.

Esquirol suggested the word lypemania from λυπεω to cause sadness and μανια madness as a substitute for melancholia. The term is expressive, but melancholy is in common usage, and is so generally received and holds so prominent a place in our nomenclature, that there is not any necessity to change it. Neither does melancholy savour of mania in the sense we have understood that form of mental disease, and I think we shall find that melancholia is a condition exactly the opposite of mania being one of depression instead of excitement, so that any term expressing mania would be inappropriate. It is a condition which not unfrequently succeeds or is secondary to mania, and it sometimes recurs alternately with it. Its marked characteristic is depression, and in this it contrasts strongly with mania. Sometimes, however, you see mixed cases, or cases in which excitement and uncontrolable impulse, from time to time, appear in the midst of the depression. Such cases have often become the subject of discussion in courts of law, and require your special consideration.

There are, perhaps, few forms of disease more painfully distressing than melancholia, and few that have afforded a wider and more fertile field for speculation. Scientific writers have indulged their fancy in this field, but imaginative authors, both prose and poetical, have found in it materials to suit sensational or popular tastes.

Among the host of authors who have made melancholy their theme, the picture drawn by Pope is as graphic as any, his description is as follows:—

> "Black melancholy sits and round her throws
> "A death-like silence, and a dread repose;
> "Her gloomy presence saddens all the scene,
> "Shades every flower and darkens every green;
> "Deepens the murmur of the falling flood,
> "And breathes a browner horror on the Wood."

We may define melancholia as a state of mental depression which engrosses the faculty of attention, and paralyses the faculty of reasoning, permitting of actions which are impulsive and beyond the control of the will.

Melancholia may be acute or chronic. The invasion of the malady is usually gradual, but it is sometimes sudden; we see, too, in this form of disease, what we constantly observe in various somatic affections, viz., that each different case is diversified by some peculiarity in the same manner as every human face differs from every other.

In illustration of the occasional suddenness of melancholia, I may mention a case, which was recently related to me. A lady, who was expecting the return of her husband from India, and anxiously watching for his arrival, was informed by a relative of his death. She uttered a loud scream, but never spoke again, and sunk into a profound melancholy, from which she never recovered.

History reports a more or less sudden invasion of fatal melancholia in the case of Queen Elizabeth, and in most records of insanity you will find similar illustrations.

Dr. Conolly records the case of a lady who lost her only son, who was her idol, by a sudden and most unexpected death. He dropped down dead in the midst of apparent health. The shock stunned and overwhelmed the unfortunate parent, and for a time grief alone occupied her. In a few weeks her state became that of deep melancholia, in which she never, in any way, alluded to her cruel bereavement, but was ever reproaching herself as sinful, unworthy to live, and deserving of eternal condemnation. She became insensible to all ordinary occurrences and affections; indifferent to her family, inactive, and silent, and attempted suicide.

To give you a clearer notion of melancholy, I will detail an ordinary acute case.

We may then consider the common attendant phenomena.

E. B., a young lady æt. 25, a member of a most united and well brought up family, was admitted into St. Luke's Hospital whilst I had charge of that establishment. At the time of her admission she said she was lost to all human feeling, that she knew the reason why her friends had brought her to the Hospital, viz: because she had attempted to commit suicide by taking laudanum, but she told me that she had so frightened herself that she had resolved not to attempt to do so again. Conversing with her on ordinary subjects she spoke remarkably well. Her information was very good and the ordinary observer might have failed to discover any insanity in her; she had for some time before been captious as regarded her food, but when she saw that another patient in the ward in which she was placed, was fed like a baby, she said she would make an effort to take her food, rather than be similarly treated, and she said she hoped she should never come into the same state as that patient. She was exceedingly neat and cleanly in her habits and dress, and employed herself generally with needlework, and for several days her state seemed to improve. She then asked for writing paper, and to my surprise exhibited a very much more marked departure from her normal condition than I expected. When in health she used to write a very clear hand, and her orthography was good, but the letter she attempted to write in the asylum occupied her the whole day, and when brought to me was on a quarter of the sheet of paper, was covered with blots, was in a small, almost illegible hand, and crossed; it was very badly spelt and full only of her moaning and complaining that she had sold herself to the devil. The next day she was very much more depressed and asked not to go out in the streets, as she had done on almost every previous day. On several occasions she asked for writing paper, and every letter was in exactly the same strain as the one I have mentioned, and after each she seemed more depressed than before.

From time to time, whenever I made occasion to converse with her, upon general subjects, I found her information quite as good as when she was admitted, and her mind clear, and her power of attention in conversation for a short time very fair. In fact her conversation was generally coherent; but she always reverted before long to her own wickedness, and to her belief that she had sold herself to the devil, that she had sinned against the Holy Ghost, that she could never be forgiven, and that in fact she had no soul, so that it was agony for her body to be alive; reasoning with her on the subject was quite out of the question; both calm, grave reasoning, and jocose and satirical interpretation were powerless to shake her fixed idea, though naturally she had a most keen sense of the ridiculous. She continued much in the same condition as when admitted for about a fortnight, when one day she put a table knife into her pocket at dinner time, the discovery of which fanned into a flame the smouldering embers that had racked this poor girl's brain for weeks and weeks. She stated that she had intended to stab herself in the night, and she was greatly distressed that her purpose had been detected and baulked. All through the night she was crying and screaming, and declared that her room was full of devils; the next night she packed her bedding into a heap and tried to get up at the window, and stated that she wished to have one more view of the London gas-lights before she died, and she told me that she had hoped she would have been able to get her head between the partitions of the iron window-frame, and so have succeeded in hanging herself: this, I am happy to say, she found impossible. For days and nights she was continuously restless, sometimes crying, sometimes screaming, sometimes lying underneath the bedstead, sometimes she turned her feet on her pillow and her head the reverse way in her bed, declaring that she was loathsome and full of devils, and that her room was full of devils, and that she herself was upside down.

She refused food, saying that it was a shame to waste it on her, and it became necessary to feed her. She also became dirty in her habits, and her condition of distress and agony of mind was the most appalling that could be witnessed.

The acute attack ran a course of a few weeks only, and she died exhausted; the *post-mortem* examination presented one general appearance, viz., that of wasting.

Her body was emaciated to the utmost degree, but I was unable to discover structural alteration in any of the viscera; they all presented the appearance of uniform wasting, they were toughish and flabby, and wanting in the firm crispness or tonicity which co-exists with healthy organs. The evidence was that of wasting, but it seemed to be the wasting of atony rather than atrophy. The brain in common with the other organs was small and wasted, I should not like to say absolutely that it was atrophic, but I may say with safety that it was atonic. It was very anæmic, and the grey matter was very pale; the cerebro-spinal fluid was considerable, but there was not a trace or sign of any inflammatory condition, and except the visible defect in size, there was nothing to show a cause or origin of the malady.

The patient had, however, a remarkable history. Insanity had shown itself in her family for four generations.

At the time she became affected, one of her sisters was in an asylum, suffering from acute mania; and I learned from my patient that the interviews which she, from time to time, had with her sister, at first impressed her most profoundly with the awfulness of insanity, but that since she had herself become affected, the dreadfulness of the idea had disappeared. I doubt, however, if we can lay much stress upon the influence of her sister's insanity as an exciting cause. She told me that her own attack was very insidious, and commenced with a simple feeling of depression. She afterwards turned her attention to religion, and by degrees she came to assume the idea that she was wicked, an idea which became exaggerated to the unbounded imaginings I have described.

You will see in this very acute case nothing, beyond the wasting, which was very definite, or upon which we can lay particular stress as to the pathology of her disease.

Another case, I would mention, is that of a man,

æt. about 40, who used to walk about almost all day and say that he was so ill that he could not recover,—that he could not get rest either day or night, and that he was sure he could never lie down again.

This man used to describe to me that his sensations were as though a thousand knives were being thrust into his brain, and from the accuracy of the description of his feelings I was strongly inclined to believe that some more than ordinary physical cause existed in his brain.

He died in Hanwell Asylum, and Dr. Rayner, the medical superintendent of Hanwell, was good enough to favour me with a note of the *post-mortem* examination.

He found the brain generally congested, and weighing 51 oz., the ventricles were not distended, but there was opacity of the membranes.

To the congestion of the brain we can attach no importance, as the vascularity of the brain varies with the mode of death. But the weight appears low, when allowance is made for the blood congesting it, and the organ must have been considerably wasted. The opacity of the membranes indicated a long continued pathological condition, but I doubt whether it indicated anything more.

His malady had a history of hereditary taint more or less perfectly indicated, and he had a son who at the present time is insane, and who at one time was under my care, the subject also of melancholia. When I first saw the father he had been suffering for three years, his disorder began with *ennui* or simple depression. In his attempts to shake it off he had taken violent exercise, and he told me that he had spent many months in climbing the mountains in Scotland, but all to no purpose, or, as it appeared to me, to the injury of himself, by exhausting a nervous organization already depressed by disease. This man never had a delusion whilst I knew him, unless the feeling that he could not lie down be regarded as a delusion. His depression and mental agony were excessive, and threw his mind so off its balance that he was restless and noisy, and at times violently excited; but that he suffered intense actual pain I have not the slightest doubt, and it makes us almost melancholy ourselves, to think that we have not yet advanced sufficiently in scientific method to be able to determine definitely the actual cause in such a typical case as the one I have detailed.

You saw at Peckham House, on Friday last, some excellent examples of melancholia. The first in a girl, who would not answer when we spoke to her. The second, an old woman with a partially healed wound in her throat which she had inflicted upon herself, the day before her admission into the asylum, in an attempt to commit suicide. The third, a woman who told us that she saw distressing visions, and heard accusing voices urging her to self-destruction.

In the incipient stage, melancholia is usually attended with a sense of great depression, a lowness of spirits which the patients try in vain to shake off; they are always fearful and in constant anticipation lest some mischief should befall them; they then begin to mope and afterwards become abstracted, moody, and sometimes sullen, and usually seek solitude, feeling incapacitated for the ordinary duties of either public or social life. Thus, as Dryden says:—

> He makes his heart a prey to black despair;
> He eats not, drinks not, sleeps not, has no use
> Of anything but thought; or, if he talks
> "Tis to himself."

While cheerless, abstracted, and seemingly idle, however, the patient's mind is not inactive, but is constantly dwelling upon his own condition, which appears to him to be more awful than that of any other living being, and with a strange mental hyperæsthesia he feels or imagines almost everything said to him, or everything he reads, to have a peculiar and special application to himself, whilst, to use Dr. Conolly's words, one dominent propensity alone is too often active, that of self-destruction.

I was once asked the question, whether a person who had attempted suicide was not often placed in an asylum in order to save a prosecution before a police magistrate? I do not think that this is so. Melancholia is a distinct and definite form of disease, which, with a little practice, you will hardly make a mistake about. Suicide is a constant morbid craving in most melancholics, but it is the result of disease, for which disease, and not because of the suicidal tendency alone, you incarcerate the patient. His attempt at suicide is from a reason vastly opposed to that of the nihilist who regards annihilation as the termination of mortality, and chooses rather to plunge at once into that annihilation than continue the struggle for existence, which is the lot of most of us. Such beings are cowards, they deliberately break the law of society, and if their attempt at self-murder fails, they deserve the punishment which the laws of society and the country impose. The melancholic, however, generally acts from a very different motive, and whilst more or less conscious that he is running counter to the law of self-conservation, nevertheless, often believes that he is obeying a mandate which he received from God. Sometimes melancholia produces a strange perversion of those sensations from which the law of self-conservation is evolved, and as a result the constant craving is for death, and if not to be met by any other means, then to be sought at the hand of suicide.

A young man was under my care in St. Luke's Hospital in whom neither hallucinations nor delusions were ever developed. At the outset of his attack he got out of health, had losses in business, and began to be low spirited; this state continued till the depression began to incapacitate him for business. He then became strongly impressed with the idea of suicide, which he afterwards described to me as an awful, overpowering, and fiery sensation which seemed to prompt him to make away with himself. He was a very amiable man, was very kind and attentive to the other patients in whose behalf he was always ready to employ himself, and he expressed the deepest sense of gratitude for his recovery which was quite complete before he left the hospital. In his case a history of insanity was not clear, but there was good evidence that he had been subjected to exhausting influence. I saw him several times after his discharge from the hospital, and he continued to improve, and his case promised to do well.

There is a point in the clinical history of melancholia particularly to be noticed, viz., that the subjects are very often better towards the end of the day, but that there is usually an accession of symptoms about bed-time when they become very restless and often sleepless. The hour of bed-time is too often merely the indicator of the aggravation of the distress. And should the patient succeed in sleeping, the distress seems to be exaggerated still more in the act of waking; fortunately however, the exaggeration generally soon subsides. But the moment of waking is worthy of very special note, as it is the time when the unhappy victim of this black despair often succeeds in accomplishing his suicidal or violent attempt; and as the brain at the waking time is with everyone in a very active condition, so unfortunably when the fixed subjective melancholy impression has become the only one occupying the mind, it often appears in its strongest and most powerful force at that moment. I have already in a former lecture spoken of the hallucinations which frequently occur in the sanest people in waking; how can we then wonder that this should be pre-eminently the moment of

"———Loathest melancholy,
"Of Cerberus and blackest midnight born
"In Stygian cave forlorn.
"Mongst horried shapes, and shrieks, and sights unholy"?

A good instance of the development of impulses in waking hallucination came under my notice very lately. A lady called upon me in great distress, because her husband, who had been under my care as a melancholic, had awakened very early in the morning, had slanderously abused her in regard to some property which had shortly before been left to her, and finally kicked her out of bed. I was convinced that it was only due to a waking hallucination, and judging from my knowledge of the patient, I believed that it would be very transient, the wife reassured, returned home and, as I expected, found her husband calm, contented and apparently having forgotten and certainly unaffected by his waking violence and excitement.

Speaking of hallucination, melancholia is the form of mental malady in which hallucinations more frequently occur than in any other, and it is only too frequently the case that such hallucinations are the incentives to the most horrible acts that a person of unsound mind can commit. Such hallucinations are usually either of sound or sight though they may be of any of the senses. The melancholic hears a voice and believes himself bound to perform its mandates. Among such cases that have been under my own observation was one, H. W., a patient who had made three attempts upon his life, and who stated that "God came to him in the night and told him to cut his throat," and he immediately proceeded to commit the unreasonable act. His attempt on that occasion was not successful. He made a second attempt which was also unsuccessful, and even while this second cut was healing he succeeded in getting his finger between the two edges of the wound and tearing out the stitches. He seemed, however, daily to improve whilst he was under my observation, and he many times said to me that he had given up his suicidal intention, but there was something so suspicious and treacherous about his appearance that I could not accept his statement without great misgiving, and at length one morning I was hurriedly summoned. He had succeeded in breaking open a box in his attendant's room, where he had been working, and extract therefrom a razor with which he had accomplished his end by cutting through the internal carotid artery. By the time I got to him he was quite dead, and I think that it was within three minutes of my having seen him alive. There can be no doubt that he must have watched this attendant most closely, and had well planned his means of procuring the weapon and simply waited his opportunity. A strange, but remarkable fact concerning the victims of hallucination and patients bent on suicide, is that they, often, will only commit the act in the way directed by the hallucination. A case was related to me of a suicidal melancholic bent on drowning himself, who through a piece of gross negligence on the part of his attendant became possessed of a razor which he took to the attendant, saying, that although the opportunity of cutting his throat had been afforded him, he had no intention of destroying himself by any other means than drowning, and which I believe he afterwards accomplished.

There was a patient in the City of London Asylum who professed that he intended to destroy himself by jumping off a precipice, and that if he once got the opportunity of getting outside the walls, he intended to make direct for Greenhithe, when he could accomplish his purpose; it was necessary to keep the strictest watch on him.

Among the extraordinary instances of hallucination that have come under my notice, was one in a gentleman, regarding whom I was engaged in a commission of lunacy.

The patient stated that he was Jesus Christ the Saviour of men, and that he was commanded by a voice from heaven to take the life of all the women on earth. He was in a private asylum, out of which he once succeeded in making his escape. Upon finding that he had accomplished the first part of his plan he forthwith rang the bell of a neighbouring house. He at once stated that he had escaped from an asylum, and asked for shelter which was afforded him. Whilst the officials of the asylum he had escaped from were being communicated with, he entered into conversation with the lady of the house, and was leading up to the special subject of his hallucination, when fortunately assistance arrived. He afterwards admitted that he would have killed the lady had he not been retaken just at that time.

It is very strange how melancholia is often associated with religious misconceptions. At one time I found on enquiring, that a large majority of the melancholy cases in St. Luke's Hospital had morbid religious impressions, and out of three cases admitted on the day I made that note, two were the subjects of morbid impressions of a religious character, both the patients declaring that they had done something wrong. One of them said that she had done all that is wicked, that she had wronged her husband, her children, her father and mother; that there was no chance of her soul being saved. The other said that she was unable to attend to her usual occupation, dreading eternal damnation for sins which the devil had tempted her to do, and that the Holy Spirit had departed, and also God's forgiveness, that she was eternally lost, and that the devil was tempting her to suicide. She constantly quoted scripture, and you will I have no doubt, frequently remark, when in practice, that melancholy patients, and especially suicidal ones, are often remarkably well versed in Holy Writ, and not only quote it in everything they say but also attempt to prove the truth of their assertions regarding themselves, by the standard of their own interpretation of the Bible.

It is certainly very remarkable that individuals who are so strongly impressed with sublime and transcendental ideas, and who have embraced the comfort that religion can give, should yet be impelled by an irresistible influence to the commission of the saddest and most awful of crimes that blot the history of the human race, and to acts the criminality of which the individuals themselves are sometimes well aware of. I knew a case of a clergyman, who placed himself in an asylum because he felt strongly tempted to commit suicide. He was otherwise rational, and reasonable. After a short time he went out on trial, but speedily returned, as the suicidal tendency returned the moment he found he was not under restraint.

Dr. Daniel Tuke quoted a case recorded by Delregne, of a patient who was opulent, stated that he was perfectly happy, and free from any cause of suffering, with the exception of one circumstance which tormented him. This was the desire, thought, or violent temptation, to cut his throat whenever he shaved himself. He felt as if he should derive from the commission of the act "*an indescribable pleasure.*" He was often obliged to throw the razor away. Many cases, as the one just quoted, are evidently the result of morbid cerebral action over which the patient has no control; but I am inclined to the belief that, sometimes the commission of suicide in melancholia is the act of logically sound reasoning on the part of the patient, who knowing full well the criminality, yet chooses to snap the thread of life rather than endure the miseries of his awful depression; the depression must indeed be something awful, when it is so insupportable that the patient rushes away from himself, and is perhaps not inaptly described in Byron's couplet :

> "E'en Satan's self with thee might dread to dwell,
> And in thy skull discern a deeper hell."

Shakespeare, who has the credit of having written something upon everything, was certainly not unacquainted with the calm and deliberate reasoning, often adopted, but too often thrown aside, by melancholics, when he penned Hamlet's celebrated speech, commencing

> "To be, or not to be,
> That is the question."

The homicidal tendency, or impulse, in melancholia, is not so common as the suicidal, but it is, nevertheless, a very common fact, and should be specially noted, as it is highly important in its medico-legal aspect. It should be noticed too, that the impulse is usually, if not always, sudden.

A gentleman, at one time an opium eater, was for some years a patient in a private asylum near London, the subject of melancholia. He used to employ himself in the garden, and although a chronic grumbler, always complaining and discontented, yet never exhibited any sign of violence, until one day suddenly, and without any provocation, he struck at the proprietor of the asylum's head with a pick-axe, and save for missing his aim, no doubt would have dashed out the brains of his intended victim. He had no spite, or ill feeling whatever against this proprietor, who had always acted most kindly towards him. He stated that the impulse was sudden and irresistible, and he was utterly unable to resist it.

A lady called upon me the other day and asked me if I could advise her regarding a friend, who had confided in her that she was unutterably miserable, and unhappy, though she had everything on earth she could desire, but that her greatest sorrow of all was an impulse urging her to destroy her children.

Such instances are by no means rare. Dr. Wilks told me of a somewhat similar one.

These cases constitute a most painful and difficult class, and the difficulty in dealing with them is especially felt in law courts, on account of the reluctance of jurors to believe in the insanity of acts committed under such circumstances; but surely no one upon consideration will attribute rationality to a mind that is perpetually prompted to commit an objectless murder.

It is extraordinary how differently the acts of suicide and homicide are regarded by jurors. Suicide, however arising, will rarely fail to find an advocacy on the ground of insanity in the mind of a jury, whilst the same jury would hardly feel inclined to believe that insanity ever existed were they called upon to decide in a case of homicide instead of suicide.

The sudden violent and homicidal impulse in melancholia was made a question in the Central Criminal Court, in the case of the Rev. Mr. Watson, and a very definite affirmative answer appears to have been given, though the jury chose rather their own, than the scientific opinion on the case before them.

But the sudden outburst in melancholia is exactly the case in which the popular notion, viz., that "ignorant people are as good judges of insanity, as the most experienced physicians," fails.

Sir W. Gull, referring to the trial of Christiana Edmunds, in the presidential address which he delivered to the Clinical Society of London in 1872, stated the case very clearly. He remarked, that the lawyer and the ordinary observer say, "the man is mad" because he has committed an act of overt insanity—a crime perhaps; but it often happens that in diseases of the brain in insanity, as in other diseases, the crime is only an indication of the sudden stress which has been laid upon a weak and diseased organ. When medical men are called upon to state their grounds for believing that insanity exists in any given case of crime, they are commonly expected to produce evidence that the disease had previously manifested itself in overt acts. But it is only when stress is laid upon a weak organ that evidence of its insufficiency is supplied.

This is the every-day experience of physicians in such coarse forms of organic disease, as mitral disease of the heart, or even some forms of peritonitis. It is the frequent experience of physicians also in cases of insanity. It is sometimes said, when insanity is discovered as the cause of a crime, that it could not have existed because it did not show itself before; but it would be more just in such cases to admit that it did exist because it had shown itself when stress was laid on the organ.

It is only a few weeks since, that I was asked to see a gentleman of about 70 years of age, whom I found extremely and acutely melancholy. He at once told me that he felt so depressed that he wished to destroy himself, and afterwards told me that upon waking that morning he felt an irresistible impulse urging him to kill his wife. He fortunately told his wife of his intention, and she immediately took steps for the protection of both the patient and herself. Here, however, was a case in which the considerable and important question of sanity, or insanity might have been raised, had the impulsive prompting been accomplished, for his insanity had not been previously observed; the miserable patient was able to express his regrets and sorrows that he should have feelings over which he had no control, and though his melancholy was so profound as to remove all doubt of unsoundness of mind from the consideration of the doctors who saw him, yet because he was able to talk rationally, some of his friends who were by him declared that he was not insane, and protested against his removal to an asylum. It is right that friends should be warned of the very grave responsibility they take upon themselves, if they refuse to let such patients be placed under proper and sufficient care and control.

A gentleman was under my care for about two years, suffering from acute melancholia, and very soon after the commencement of the attack though depressed to the utmost, constantly exhibited outbursts of violence, the impulse to which he assured me was irresistible; one morning at breakfast, he threw his bread at his wife, on another occasion he took up from the table some knives and threw them with violence against the door, on another occasion he broke a looking-glass with his fist, and he committed many similar acts, all of which he detailed to me with sorrow and contrition. I was convinced that they were irresistible, from my acquaintance with all the details of the case; yet, during the whole of the time that his malady lasted, had a stranger, unacquainted with any of the facts of insanity, visited the house, he would, I think, most certainly have failed to discover any insanity in the patient at all. At one time, however, I was nearly driven to despair of the gentleman's life, so profound, depressing, and exhausting did his malady become.

Many, more or less chronic, melancholics are, often, unable to restrain their feelings, and are constantly telling you of their imaginary sufferings, following you about and begging for advice.

An extreme case in St Luke's Hospital, used to watch for me daily, and follow me about the ward constantly asking me what she was to do, whether I would forgive her, and if sometimes I tried to encourage her by an answer in the affirmative, she would reply, "it is no use, for if you say you forgive me I feel just the same as I did before"; this I would also bring under your notice as showing you how material a thing that which we call mind is, and that the conditions of its diseases are material conditions. It is substantially true that my forgiveness of this girl's imaginary wrongs, which she believed I had the power to forgive, was useless, and that she did feel the same awful depression, she felt before; the subjective idea of forgiveness not changing the condition of the brain, or re-establishing its tone. But although a great number of cases are unable to repress their lamentations, yet some, at times, have the power of self-control most singularly marked in this respect, and thus it is that the profession of having relinquished suicidal and other tendencies formerly exhibited, must be received with the greatest amount of caution. Melancholy patients will often appear for a long time quite well, and not revert for months together to any of their melancholic feelings, in order to induce you to believe in their recovery, but the moment they obtain their liberty they commit suicide. I knew of one very sad case, of a gentleman, who was liberated from an asylum apparently quite well, but, shortly afterwards, blew his brains out with a pistol. Among the instances of self-control that have come prominently under my notice, was the case, most of the particulars of which I have already detailed, viz., the gentleman who believed his mission to be the slaughtering of all the women on the face of the earth. Many times before the commission, he had told me that he was the Lord Jesus Christ, but on the morning of the day upon which the inquiry was held I found it almost impossible to discover his delusion.

He did not commit himself on any topic of conversation addressed to him, until he was asked whether he was not the Lord Jesus Christ, when he immediately affirmed that such was the fact.

Dr. Daniel Tuke mentions a most interesting case of a gentleman who was the prey of melancholy, and who could never restrain expressing his gloomy feeling before his wife, and could not attend in the least to his family affairs, but who one day paid a visit to the poet Southey. Subsequently the poet, who was cognizant of his guest's mental imfirmities, expressed to the patient's wife the satisfaction he experienced in seeing him so well, and added, he never knew him reason more clearly. On the wife repeating this to her husband, he exclaimed, "why, you know, I could not think of showing my weakness before *him*."

Although many melancholics are for a time free from delusions yet, mostly, they at length come to the assumption of one or more fixed ideas, such as we saw in the religious cases I have detailed, and some most excellent illustrations were afforded to us at Peckham House on the occasion of our visit on Friday week last. You remember well the case of a woman who stood against the door and told us that she feared for the safety of her children, and that she had never done her duty by her children, or her husband, and during the whole of our interview she was constantly in tears. This woman, labouring under the sad and distressing idea, had in her own home, before her removal, constantly placed in cupboards and various parts of her house, food in superabundant quantity, as she said for her children, and the evidence is that she had always been an exemplary mother, and a fond and affectionate wife, and the excitation of her attack seems to have been the loss of her husband. You might have remarked that although she spoke well, and apparently rationally, she was crying during almost the whole of our interview, and this was a somewhat bad symptom. The lacrymose melancholia as it is called, is to be found in perhaps not a large number of cases, but the characteristic is always marked and prominent, and in an asylum you generally see one or two cases; the patients of this class require special attention, because they are always suicidal. Though they will speak nicely, and affectionately,—and with a clearness of reasoning truly remarkable, when it does not concern their own particular fancy,—and though they will give expression to the sublimest ideas of religion, morality, and refinement of feeling, they will go away from you and immediately afterwards destroy themselves.

In St. Luke's Hospital was a case very similar to the one you saw at Peckham, but of longer standing, the patient was constantly crying, fearing some harm had come to her children, all of whom were doing well; she, poor creature, was constantly watching for an opportunity to commit suicide, at which she had made several attempts, and I found treatment to palliate, soothe, and calm her constantly called for.

Pathology.

The evidence we may deduce regarding the brain, from physical conditions in melancholia is very valuable, for it is often that you can trace some connection between the physical conditions and the mental phenomena. The predisposing cause is mostly hereditary, but without some physical excitant, or perhaps it should be called depressing cause, the melancholy probably does not occur.

A gentleman called at St. Luke's Hospital one day, and described to me his melancholy feelings. I saw him a few days afterwards and his depression had increased, and a few days after that it had increased still more; he was sleepless and restless, complained of headache, and he became so irritable that his children were sent from home to save him from their noise, but his depression increased, and became so burdensome to him that he requested admission into St. Luke's, lest he should, as he said he feared, lose control over himself. He had become weak and feverish, had a furred tongue and confined bowels, and frequent aching of the head.

At the commencement of his malady he was usually better in the evening, provided he had partaken of a little food and wine. His headache then disappeared, and he used to be able to change the current of his thoughts, from the feeling of gloom and despair to some light amusement, as a game of cards ; but this promise of improvement was not fulfilled, and he became weaker and more irritable, and one morning he wantonly put into his mouth, and swallowed, a small piece of paper he had in his hand, this became the turning point of his attack ; he imagined that the paper had stuck in his œsophagus, and he referred to a spot in which he felt pain, no doubt of dyspeptic origin, in which he declared that the paper had stuck ; from that moment the depression increased, the exhaustion continued, the anxiety was not moderated, and finally the mind gave way. His circulation became depressed to an alarming degree, his limbs and extremities became cold, and the skin of one of his fingers sloughed. He expressed his intention of commiting suicide, and it was necessary to keep the strictest watch upon him, nevertheless, he was able withal to speak and converse from first to last, with perfect coherence and rationality upon any subject, but that of his own malady and depression.

This was a second attack, excited by over-straining of his mind in his business. He had allowed himself to become over anxious, and at length over worked. He got headache and depression, and an uncomfortable restlessness, and the general symptoms of brain deterioration in the first instance ; he then became captious as to his food, refusing first one thing then another, till at last he became exhausted from want of food, his condition went on from bad to worse, and he very nearly died from exhaustion, but under the treatment of feeding and stimulants he quite recovered, and as his general nutrition was restored, his morbid impression regarding the piece of paper entirely faded away.

He had a clear hereditary history of insanity, and several of his children have shown symptoms of brain imperfection.

Sometimes from disorder of the intestinal canal the patients will refer all their complaint to some part of the tract. You remember the patient the other day, who told us that she had no intestinal tract, that her inside was a cavity, through which the food dropped unaltered, and that she thought she was weakened by it, and that it had become so through the possession of an evil spirit which had entered into her inside; you may remember too how well she spoke, and the shrewd observations she made about other patients in the ward. She has, I believe, recovered from the intestinal derangement from which she suffered, but she continues very weak, and with her weakness the morbid fancy persists. I once had under my care a patient whose case I published,* who from confinement of the bowels came to assume the same idea, viz., that he had no bowels at all; relief was soon afforded him, but he persisted for some time in the notion that he had no bowels, and afterwards declared, because he happened to soil his clothing and his bed one day, that his belly was a void

---

* Matter and Force in Relation to Mental and Cerebral Phenomena. *Journal of Mental Science*, July, 1869.

through which everything he put into his mouth arpidly passed. Among the curious cases of extraordinary morbid fancy in melancholia that have come under my observation, were two women, both of whom had abdominal tumours. One of them declared that she had a weasel in her inside, the other that she had a little dog in her belly.

A very remarkable form of delusion which sometimes accompanies melancholia, is that, in which the patient imagines that he, or she, has become transformed into one of the lower animals which he or she apes or imitates, the patient often assuming the attitude of all fours. I had a most interesting case of this kind in St. Luke's Hospital, the exact form being an instance of what is called "canine madness"; the patient, a married woman, who believed that she had become transformed into a dog, would constantly bark, would run on all-fours unless checked, and unless carefully watched at meal times would feed herself as does a dog, by applying her mouth to her food. Curiously enough, she was always willing to converse, and would talk of her husband and her friends (she had not any children), she would then fall back into melancholy complaining bemoaning the sadness of her condition, and "the awfulness of the scourge of a reasonable human creature being converted into a dog."

Hysteria sometimes assumes a sort of dog mimicry, with a cough very like a dog's bark.

I shall have occasion, by and by, to make some remarks upon hysteria, so need only note here, that the melancholy of the woman I have spoken of was extreme, and her canine impression was that of a strong unreasoning delusion.

A certain amount of confusion of terms has arisen from the fact that hydrophobia has been called " canine madness," from the fact that it is usually communicated by the bite of a dog suffering from that disease. Several animals, however, can communicate hydrophobia. Hydrophobia is a fatal disease, it is attended with mental symptoms, and has its origin in blood poisoning. In its course the kidneys become acutely affected. Happily the disease is rare. Madness in dogs is not always due to hydrophobia, in fact, hydrophobia is the exception. The common madness in dogs is rabies, a different disease altogether, and one not communicable to man. I recommend to your notice an important paper on this subject, by Dr. E. P. Philpots of Poole, Dorset, published in the *British Medical Journal*, of March 8th, 1873.

Strong and prominent as is the contrast of melancholia with mania in its manifestation, I am inclined to the belief that the actual cerebral conditions in the two forms of disease are not so very opposite. The ultimate cause to which you may trace back all cases of both forms of mental disease, is imperfect nutrition. In melancholia, however, the imperfection appears more general than in mania, and the evidence does not point to cerebral hyperæmia as in mania, except in the extreme and excited cases. In mania it would appear as if a portion of the brain only were primarily affected, while the remainder released from control, rapidly and excitedly performs its function, the condition being compared to the heart in fever, (as I spoke of in a former lecture). Melancholia on the other hand is a condition of general mal-nutrition, and is not marked by any undue excitation, or any particular loss of control, the condition being that of general atony, and consequent imperfection of function. A delusion, when it occurs in melancholia, does so because there is not energy enough to correct a false impression, or a mistake of the sense.

The general state is very like the general condition of weakness which follows any severe form of disease, and in fact, the so-called lowness or depression of spirits which follows severe illness, as fever, is due to an analogous condition, operating, however, in a minor degree. The patient in fact, may not inaptly be compared to a plant, the leaves of which are drooping for want of water. The disease may be imagined in the idea of withering, an idea always more or less melancholy, but one sustained in its sequel, viz., if appropriately treated in time, recovery in most cases will crown with success the restorative efforts used. In its absolute pathology, the cerebral hyperæmia which attends mania, appears to be absent in melancholia, and the differentiation in the symtoms of the two forms of disease appear to be in the one case, *i.e.* in melancholia, actual deprivation of nourishment and in the other arrest of normal nutrition, consequent upon the paralysis of the vasi motor nerves which is always attendant upon congestion.

Coarse morbid changes are not usually found in the brain and membranes of those who have died in melancholia, and when they do occur they resemble very much the changes which are found after death in mania. But as a rule, except the evidence of atrophy, it is not common to find anything which strikes you as being important.

Thus you may find the membranes thickened and opaque, but I have found them exceptionally thin; and I feel convinced that if you do find coarse and visible changes, none of them are the direct cause of the melancholia, except in so far as that they may arrest nutrition, and I doubt if you can attribute, as a result of the melancholia, any of the coarse conditions which you may find within the skulls of melancholy patients.

There are, however, certain microscopical conditions which I have found and which you may judge of for yourselves from the specimens on the table. The cases were characteristic cases of chronic melancholy, and the specimens marked Nos. 17 and 18 were from the brain of a patient who poisoned himself with hydrochloric acid.

You will notice that these sections contain very wide perivascular spaces and you see the vessels lying in these spaces, some of them surrounded by sheaths, others free, and distributed on the vessels and upon the sheaths you notice that there are granules of blood pigment or hæmatozin. I shall, further on, show you a somewhat similar condition in the case of the disease commonly called general paralysis of the insane,—the changes in which neurosis were first minutely described by Dr. Lockhart Clarke who investigated the appearances detailed by Weld, Sankey, and Rokitansky.

There may be a difference, however, in the two pathologies—a difference perhaps of degree, for in the case of *Progressive Paralytic Insanity* the vessels are corrugated and contorted into kinks and knots, and the hæmatozin grains are in great quantity; whilst in melancholia the vessels remain more or less straight, and the hæmatozin, though in considerable quantity, is not nearly so profuse as in paralytic insanity.

Among the sections you will find some from the gyrus fornicatus, one in particular in which a vessel has been lifted out of the sheath and turned back. You will see the cells of the grey matter very clearly in the specimen marked No. 19, and you will notice that their outlines are indistinct and not sharply defined; they are in fact wasted, but they do not appear granular like the cells of general paralysis, as described by Dr. Lockhart Clarke. I have, however, found the same conditions in wasted brains which have never given any indication of insanity at all.

The vessels are not materially changed in their structure, unless they have partaken in the general wasting. They do not, however, appear to be fatty.

I cannot at the present time give you a satisfactory explanation of the presence of the hæmatozin. Many more facts are wanted in order to estimate the influence of attendant conditions.

There may be, and I am inclined to believe generally is, a premonitory stage in melancholia, during which there is more or less cerebral hyperæmia. The patient does not complain; he is not bad enough to seek medical aid, but he feels a constant low aching of his head, attended with a sensation of heat, a condition which is attendant upon a low degree of vascular paralysis and impaired cerebral circulation. It may be that at this time corpuscles of blood escape from the capillaries, and, undergoing the ordinary changes which occur in effused blood, leave the common evidence of their visit, viz. granules of pigment in the sheaths and in the canals of the vessels.

The prominent fact in the pathology of melancholia, however, is that of wasting, which is testified to by the condition of the cells, and of the white substance, which confirms the opinion of mal-nutrition, which will strongly impress you as you watch melancholy patients during life.

LECTURE VII.

Etiology of Melancholia.

Heart Disease and Melancholia—Post-mortem Condition of Various Viscera—Syphilis—Diagnosis of Melancholia—Treatment of Melancholia—Rational Principles—Surroundings—Food—Feeding—Mechanical Feeding—Stomach Pump—Nasal Tubes—Nose Feeding—Drugs—General Paralysis of the Insane.

As with mania so with melancholia; the predisposition to the malady usually dates from a former generation, and the essential exciting cause is a depressed state of vitality or an over-taxed brain.

That a predisposition may be set up, is probable, from the history of such cases as those associated with rheumatism and heart disease, in which hereditary history is wanting. It is nevertheless open to question, whether the depressing effects of anæmia and rheumatism or the impeded supply of blood to the head in cardiac disease would produce melancholia, unless a predisposition already existed.

It may be argued that if melancholia can be set up by rheumatism or the invasion of heart disease, it ought *a priori* frequently to follow these conditions. But it may also be advanced that as melancholia is found to follow rheumatism and heart disease, and that as a large percentage of recent melancholics either have morbus cordis, or have lately recovered from acute rheumatism, *a fortiori* the rheumatism or heart disease stand in the relation of cause. I am inclined to believe that a mild form of melancholia almost always follows acute rheumatism or accompanies early heart disease. In both cases you almost always see great depression of spirits and this may readily overstep the boundary line of sanity, or upset the balance of normal and healthy feeling and thought.

Morbus Cordis is sometimes the cause of melancholia, by its direct influence in diminishing the supply of blood to the brain.

In my experience mitral imperfection sets up a transient melancholia, but aortic disease produces a more permanent form of the disorder, and as already stated, a transient melancholia often follows in the train of acute rheumatism, probably from the united effect of the attendant anæmia, and the associated cardiac imperfection.

I may here remark that opinions vary very much as to the influence of the heart upon mental states. Dr. Sutherland stated in his Croonian lectures, that of 40 patients examined in St. Luke's Hospital in 1853—56, the heart was healthy in only eight cases. Dr. Blandford* does not connect the morbid appearances of heart found on the *post-mortem* table, with the outbreak of insanity, and considers that they are the result of long continued violent and irregular action of the organ during many years.

Dr. Maudsley† remarks that "observers, agreed as to the frequency of the occurrence,"—of diseases of the heart—"differ as to the proportion of cases in

* *Lect.* p. 83. † *Loc. cit.* p. 470.

which they are found: Esquirol found them in one fifteenth of his melancholic patients, Webster in one eighth, Bayle in one sixth, Calmeïl and Thore in nearly one third. The most reliable observations of late years tend to lessen the exaggerated proportion commonly assumed; out of 602 *post-mortem* examinations in the Vienna Asylum, affections of the heart were met with in about one eighth of the cases; and in some of these the disease was very slight." But *post-mortem* statistics are in this instance falacious, and afford the most imperfect basis from which to draw a conclusion in this matter, because a large proportion of the melancholics whose malady is traceable to heart disease recover. Statistics on this question must be based upon accurate observations made upon patients at the time of their entrance into asylums, and these should, if possible, be supplemented by observations made upon the less severe cases which are frequently seen in private practice. I have not any accurate statistics to place before you, but my observations led me to believe that if $12\frac{1}{2}$ per cent. be the average of heart diseases found in melancholics after death, that the per centage is very much higher when considered in regard to cause, and the number of transient cases and recoveries counted in the average.

In regard to the viscera generally, changes, bearing more or less, by their exhaustive influence, upon the melancholia, are found. Though the melancholia may be considered the result of the exhaustive influence of the depraved organs on the brain.

The *post-mortem* table, constantly exhibits the morbid conditions of heart and lungs giving rise to the irregular circulation which as we have seen is frequently a prominent feature in the symptoms which appeared during the life time of the patient. But the disorder of function of many of the various viscera, such as the stomach, which produces so much disturbance during life, leaves little or no trace, or footstep behind it, and from the evidence of the *post-mortem* table, there often is nothing to show that such disorder has existed.

The complication of melancholia which is very fatal, is a form of lobular pneumonia, running on to gangrene. The condition is not peculiar to melancholia, it is to be found in other conditions of depressed vital energy, and appears to be not an uncommon form of pathology found in work-houses. I have not often seen it in the *post-mortem* room of this hospital though cases occasionally are brought in. You will find the pneumonia or the gangrene affecting the lungs in isolated lobules, and the surrounding tissue wasted and ill nourished.

A possible cause is syphilis, though the form of mental disease which most commonly has its origin in syphilis, is probably dementia or paralytic dementia.

I not long since saw a very perfect case of melancholia in which the history pointed prominently to syphilis; the patient had, however, been drenched with physic and some of his depression was in all probability due to the drugs he had swallowed. His syphilis may, however, have set up a change in the nutrition of the brain's vessels and so impaired the nutrition of the brain itself.

That syphilis will set up change in the vessels of the brain we can have no doubt: and though the exact pathological condition may have been but imperfectly demonstrated, nevertheless we are able to see a change, at all events we see an increase of tissue. I have placed the only section of a syphilitic cerebral vessel I have under a microscope for your inspection, but it exhibits the muscular and fibrous coats too perfectly, and I am afraid that it shows too much to draw any conclusion from.

Diagnosis.

The only form of mental disease with which I think you may confound melancholia is acute dementia, in some cases of which the patients refuse to speak. However, with a little patience you will generally get the melancholic to give some expression to his feelings, whereas in acute dementia you rarely will so succeed: again, in acute dementia you will not find the appearance of settled gloom, you see in the silent melancholic: the history also may help you, but you will not find the same degree of restlessness and sleeplessness in the acutely demented as in the melancholic.

The mental phases of the acutely melancholic will frequently vary. At one time, the patient will be obstinately stolid and mute in your presence, or will turn from you and hide his face in a corner;

at another time he will pace the floor, throw his arms up in an attitude of despair, and exhibit signs of the utmost anguish and mental pain; sometimes these patients will commence to bemoan their fate audibly, as soon as they believe themselves to be alone, and you may hear their mournful complaining almost before you have closed their door. The acutely demented, on the other hand, will sit, stare, seem willing to listen, but unable to answer, and appear perfectly emotionless and insensible of either external or reflective influence.

I mentioned in an earlier case that the patients' handwriting and letters were sometimes characteristic. The reason of this being that when they sit down to write, their thoughts wander away to their melancholy, and occupy them too much to allow them to indite. I hand round for your inspection the scribble of a melancholic lady, who in health was able to write easily, intelligently, and well; it is very different, however, from the agraphia of paralytic patients, of which we shall have to speak by and by, and it contrasts well with this letter which I also hand round, and which was written by a lady suffering from semi-acute mania. It is well written, fluent, well expressed, and though absurd, it is coherent.

Treatment of Melancholia.

We must now consider some details of treatment, many of which, like those with which I concluded the subject of mania, are applicable to the whole range of mental diseases; but I have chosen this place for the remarks I am now about to make, because a knowledge of many of the points I shall comment upon is essential to success in the treatment of melancholy.

The medicines and remedies which have been proposed for the treatment of melancholy, are as endless, superstitious and poetical as the descriptions of the disease itself. Gold, precious stones, witch-craft, herbs, and all kinds of absurdities have been vaunted, and their virtues praised by numberless writers; and among the numerous, fanciful, superstitious, amusing, and impracticable remedies detailed by Burton,* for the cure of melancholy, perhaps his paragraph on gold is the most amusing, and at the same time as fanciful as any. He says —" Most men say as much of gold and some other minerals as for precious stones. And he quotes Erastus, as maintaining an opposite opinion, who, he says, " confesseth of gold, that it makes the heart merry, but in no other sense, but as it is in a miser's chest," *at mihi plaudo simul ac nummos contemplor in arcâ*, as he said in the poet, so it revives the spirits and is an excellent recipe against melancholy.

> For gold in physic is a cordial,
> Therefore he loveth gold in special." (Chaucer.)

Auram Potabile† he discommends and inveighs

---

* *Anatomy of Melancholy.*

† Some amusing recipes for Auram Potabile you will find

against it, by reason of the corrosive waters which are used in it. *Aurum non aurum. Noxium ob aquas rodentes,* and concludes their philosophical stones and potable gold "to be no better than poison," a mere imposture, a *non ens;* dug out of that broody hill, belike this golden stone is, *ubi nascetur ridiculus mus.*"

In our day, I do not think that any one will be tempted to prescribe aurum potabile, or dissolved pearls, and one thing is very certain, gold will not buy physic that will restore reason to the melancholic, any more than it will purchase balm or nepenthe to soothe or heal a broken heart. Yet we cannot altogether,

"Throw physic to the dogs,"

for it is possible to minister even to a mind diseased, and it is quite certain that the maladies or diseases of the mind, are quite as amenable to treatment as any other class of disease, which flesh is heir to.

By early and judicious treatment you will save the reason, or have the satisfaction of seeing the patients restored to his ordinary mind, in the majority of the recent cases that come under your notice, whilst delay or uncheering influences may consign the poor sufferer for life, to hopeless and irrecoverable insanity.

in "Choice and Experimental Receipts in *Physic* and *Chirurgery,* collected by the Honourable and truly learned Sir Kenelm Digby, Knight, Chancellor to her Majesty the Queen Mother. London, 1668."

The essence of our treatment, must, however, be reason, and we must treat every case upon rational principles. Tentative treatment will most certainly fail. In melancholia as in mania we must remember that the evidence we have is that of malnutrition of nerve substance, and exhaustion of nerve power, an exhaustion too which has a tendency to become maintained or permanent by the depression or excitation under which the poor sufferer is labouring. If we allow such patients to be up and out of bed, we are only allowing them to expend the remaining potential energy which their impaired brains possess.

Activity is natural to the brain of everybody when awake, and particularly when up or out of bed, and so, in melancholia as in mania, in order to husband nervous force, our first care in every recent case, should be to put the patient into bed and keep him there, strictly watched if necessary.

The most restless and noisy patient will in a short time, perhaps a few hours, at all events a few days keep in bed without attempting to get up; and I may repeat here that a little kind and gentle persuasion will often calm the most pitiably distressed, or render tractable the most obstinate patient you may have to deal with. You must nevertheless be firm, for if your patients discover that you waver in the slightest degree they will gain the mastery. If you are weak and undetermined, they are clever enough to see your weakness and to take advantage of it.

A suicidal patient at one time under my care

discovered that her attendant wavered a little in insisting upon her food being taken, and by degrees she lessened the quantity she took, and increased the quantity she left, until she at last took none at all, and I was obliged to resort to mechanical feeding, before I could again induce her to feed herself.

The influence of moral control is not as great in melancholia as in mania. The presence of a sane person does not afford the same support to the wretched melancholic, sunken in the slough of despair, as it does to the excited lunatic who is able to derive a species of pleasure from the balance which a sane mind gives to his own. Nevertheless, association is essential to the melancholic, who, both day and night should be with other people. One value at least of association is, that it reduces to a minimum the opportunities of suicide. The melancholic usually seeks solitude wherein to put an end to his existence, and I believe that open attempts at suicide are never determined and rarely fatal. In the day-time a person will hardly attempt suicide in the presence of other people, whether sane or insane; and at night-time any attempt at self-violence, will, in all probability, awaken one or other of those who may be sleeping in the same room. Whatever advantage is to be gained from association in melancholia, must be taken. The patients will seek seclusion and try to get out of observation; and you must remember this and impress it upon the servants and attendants under you, for under no circumstances should a melancholic patient be allowed to be alone. Sometimes, however, you may find a certain degree of isolation of value. The noise of association may be painful and wearisome to the patient, and a separate room will often afford comfort, and the watchful eye of one reliable attendant will cause less distress than the associated ward; and a study of these circumstances will help you in your endeavours to extenuate the depressing surroundings and mitigate the sufferings of your charge.

I was requested to see a lady during last winter, who was reported to be suicidal, and I called upon her at a lodging where she had been for some months. The house was low lying and damp, and the room she occupied was cheerless and dark, and the patient was miserable and melancholy and the prey of delusions and unutterable despair. Her bedroom was overcrowded and more depressing than her sitting room; and the house was full of children whose constant screaming and crying increased to a degree beyond conception, the distress and agony of the lady. Convinced that living under these influences her melancholy would of necessity increase, I recommended her removal to a quieter and more cheerful scene, under the strict surveillance of a trustworthy servant. This was done, and as I found that the patient had been refusing her food, I insisted upon certain quantities being taken with regularity; and I had the satisfaction of seeing her cheerful and free from delusions, and greatly improved in general health within one month of the time when the change was made.

The second essential of successful treatment is food. You will usually find that melancholics have a distaste for food, or perhaps from obstinacy, perhaps from a determination to starve themselves to death they will refuse it. Upon gentle persuasion they may sometimes be induced to take it; more often you will find that it is necessary to feed them with a spoon like you would feed a baby. If such attempts fail, forcible mechanical feeding must be promptly resorted to. The old system of feeding was with the stomach pump, and to my mind it is still the best, it is undoubtedly the easiest means of insuring that the patient has a sufficient quantity; and it is a mode very easy of application; it is also more expeditious and safer than any other plan.

You have of course to secure the patient, and hold him so that he cannot struggle, and the only method if he is refractory, is to place him in a high backed arm chair, (a Windsor chair with arms is the best) to which he must be bound with round towels. His arms and legs must first be secured, and then his body; his chest must be allowed to expand freely in respiration, but he must be held so firmly that he may be incapable of any sudden jerk or motion.

Having secured your patient, you have next to overcome the obstacle of his closely clenched teeth. This separation requires nice manipulation, otherwise you may break the patient's teeth, an accident which often has happened with the old fashioned wedge. The best form of wedge, and one which you may use without risk of harm, is an ingenious wedge-shaped screw which was invented by Dr. Henry Stevens, one of my predecessors at St. Luke's; it should be plated, with nickel or silver, and it may be insinuated without force, and by gentle screwing it will gradually open the jaws; as soon as you have the jaws open the gag must be inserted between them and it may be held in its place by an attendant standing behind the patient. The introduction of the tube greased with castor oil and the remainder of the mechanical processes of filling and emptying the syringe I need hardly detail, you will learn them in the surgery, and a little practice will make any one of you adept at the work, but I recommend you to use a tube of large rather than of small diameter, as I think it passes more readily, and its passage is attended with less risk.

The use of the stomach pump may appear to the mind of many of you a very barbarous mode of procedure. But its great advantage is that it is summary and more speedy than any other plan of feeding, and that it does not exhaust the patient as other systems of forcible feeding often do. The performance too, though very formidable in appearance, has not only the justification but also the recommendation in that it is safe. There are cases on record in which it is said that the operator has pumped the food into the patients chest, but such stories are apocryphal, for the tube of the stomach pump, cannot pass the rima glottidis or into the larynx without almost instantly suffocating the patient, and if by some unaccountable bungling the tube should have been forced through the walls of the æsophagus, the accident would immediately become manifest.

A good plan of feeding, which was warmly advocated by the late Mr. Moore whilst surgeon of St. Luke's, has been often adopted with success. The method consists in passing a tube through one of the patient's nostrils, and pouring the food through the tube by means of a funnel, thus the food trickles to the back of the pharynx and is involuntarily swallowed. This system has the recommendation of ease in its application, and saves the risk of breaking the teeth which the most skilful application of the gag, sometimes is unable to avoid, especially if the teeth are decayed, but from the necessary narrowness of the tube, very fluid kinds of food only can pass it, and this is a disadvantage. Dr. Harrington Tuke advocates the passing of the nasal tube through the nostril into the patient's stomach, but this procedure increases the risk attending the performance, and is a work of supererogation, for the fluid ocne reaching the back of the pharynx must of necessity be swallowed involuntarily.

You will find a paper in the *Lancet*, March 20th and 27th, 1869, by Dr. Anderson Moxey, recommending another method of nose-feeding. He inserts a small funnel into one nostril and then pours fluid food through it. But I think that there are some serious objections to the plan, not the least of which is the preliminary struggle; then there is a risk of choking the patient, because his head has to be thrown very far back, and easy deglutition is thereby likely to be interfered with, especially when the patient is determined to make efforts to get the food into the mouth and spit it out; and thirdly, the difficulty of administering alimentary substances of sufficient consistence to be nutritious.

There is a method of feeding with a spoon over the tongue which I frequently adopted in St. Luke's Hospital; it is very simple in its appliance and the necessary apparatus is always at hand; but it is sometimes troublesome and difficult to apply, and ought not to be attempted if the patient is weak on account of the strong resistance to be overcome and the exhaustion which often ensues. The method is advocated by Dr. Williams of the Sussex Asylum, and he described its detail in the *Journal of Mental Science* in Oct. 1864. An attendant should hold the patient's head on his or her knees, and separate the lips by the insertion of the little fingers into the mouth at its angles. You can then introduce a spoon between the teeth, and you must pass it far back in the mouth and so get command over the tongue. With another spoon or with a feeder, you can then pour the food little by little into the mouth. If the spoon be over the tongue the patient cannot spit the food out, and in a little time will swallow it. Sometimes it is necessary to pinch his nose, when in his attempt at respiration the food reaches the pharynx and so goes down. Of course you must take care not to put too much into the mouth at once, nor to put a second supply in until the first has been swallowed. But there is one great objection to this plan, it often involves a great struggle with the patient, and this is attended with danger. On one occasion I witnessed a patient die at the end of such a struggle. The patient was secured, the spoon introduced into the mouth, and a spoonful of food was administered and swallowed, but the patient showed signs of collapse and died within a few minutes. The cause of death was from the failure of the heart which was very large and fatty, and unfit to bear any strain such as the feeding struggle entailed. There are many cases, however, in which the method may be pursued without risk, and in such cases, when it can be readily carried out by intelligent attendants, under your supervision, it is a plan which may be usefully adopted. At least four, and often six, attendants are required to hold the patient. The common plan adopted is to place the patient on a bed (Dr. Williams says on a mattress on the floor) with his or her head resting on the knees of an attendant, who is also to hold it; a second, and if necessary a third, attendant must secure the patient's knees, two other attendants must take an arm each, and it is useful to have a sixth to assist the operator, and hand spoons or food as required.

The food to be given by mechanical means must be of the most nutritious character. Meat minced or ground to a pulp, and mixed with beef-tea, to which may be added brandy or wine, is the best aliment you can employ. You may mix with it bread-crumbs, arrowroot, Indian corn flour, or any other farinaceous food you please, and the whole quantity at each feeding should be about a pint, at all events not over a quart. You may alternate the meat diet with raw eggs beaten up with milk and brandy, or with rum and milk, or with cream, or any other nutritious fluid you can obtain.

Many patients get to like mechanical feeding, and refuse to take nourishment in any other way.

There was a patient in St. Luke's Hospital who was fed night and morning for nine consecutive months with the stomach pump.

The third essential of your treatment must be the regulation of the excretory organs, and first the bowels must be kept regularly cleared. You will observe that in melancholic, as well as in maniacal, as indeed in all insane patients, the alimentary canal, and all the abdominal organs are sluggish, and you must direct special attention to them, and you will often find a good purge work marvels, especially if the bowels have not been open for two or three days. But you must be careful not to depress your melancholy patients with purgatives, nevertheless you must give moderately large doses.

In St. Luke's I used to prescribe the compound senna mixture of the Pharmacopœa, and found it an efficacious purgative, but it is not pleasant physic. A dose of castor oil, or a castor oil draught made with castor oil and tincture of rhubarb and flavoured with oil of cinnamon, will often do its work well; but an occasional mercurial purge is often the most valuable, and may be administered in the form of calomel, blue pill, or grey powder, with colocynth and henbane. You may sometimes find Freidrichshaller Bitterwasser very good physic, it has lately been much recommended, but it is very nasty. Enemata may be used daily if necessary, and sometimes are of much service.

In melancholia baths are of the utmost importance, and will often give the patients great comfort. At first it is well to order a warm bath every morning, and, if the patients are sleepless, a repetition at night, afterwards shower-baths, at first warm, and then gradually cooled down to tepid, but never quite cold, are of the greatest possible value. In hot summer weather water that has been standing for some time in tanks is rarely quite cold, and may be used without the addition of warm; but water just pumped from a well is too cold and would act as a shock to the patient, it should therefore not be employed.

I would lay stress on this point; absolutely cold baths in melancholia should altogether be avoided; the *vis vitæ* of melancholy patients is already low, and they have not sufficient vitality to bear the shock which a cold bath produces.

A very good form of bath is an ordinary warm bath, 90 to 100 F., with a cold douche or a bag of ice to the head; these are very comforting, and a patient will often sleep quietly for hours after them. If the patient is the subject of amenorrhœa you will often find that hip baths administered nightly are more advantageous than ordinary baths.

The remarks I made upon the subject of narcotics and sedatives under the head of the treatment of mania apply with equal or even more force to melancholia; but observers are not by any means agreed on this point, and you will find that opium is recommended in the text-books upon mental diseases. Dr. Blandford recommends the use of bimeconate of morphia in acute melancholia, he also recommends the use of subcutaneous injections of solutions of morphia. The latter I have found useful when the excitement has been great, but I am very strongly of opinion that more than three or four doses should not be given, and when it is employed it ought to be given only as a temporary expedient to allay excitement. Dr. Blandford seems to regard opium as exerting a healing influence, and he has found it particularly of service in sub-acute melancholia. The observation does not accord with my own experience; any temporary relief which opium may afford to the melancholic patient is I believe always followed by increased depression, and I am convinced that ultimately in all cases it is injurious. If a sleeping draught is necessary, give chloral, or chloral and bromide of potassium together. If you cannot get them taken by the mouth, inject them into the rectum. But you will benefit your patient more if you can induce sleep with food and stimulants, and by the avoidance, if possible, of narcotics altogether; of course if their exhibition is called for in the treatment of surgical or any local painful affections you cannot help but give them, their use be discontinued as soon as possible.

The patients all require stimulants, and having regard to the imperfect or depressed circulation so common in melancholia, you must give stimulating and tonic drugs. Iron, quinine, cod liver oil, mineral acids may all be found of service, and must be given whenever indicated, or the patients are in a condition to bear them. Mineral waters, in particular Vichy, and Carlsbad, may also prove useful in correcting and stimulating the secretory and excretory organs.

Special conditions must of course be specially considered, and you will find menorrhagia, amenorrhœa, leucorrhœa, gonorrhœa, secondary and tertiary syphilis, and other constitutional and local diseases, in melancholy patients, calling for immediate and appropriate treatment. Of course you will treat these upon the ordinary principles of medicine, and to the best of your judgment, bearing in mind that your one great object is to husband the nervous power of the patient, and to endeavour to restore tone to his weak and depressed brain cells.

Cardiac affections, as indeed all organic diseases, you can only treat on general principles, but when they are the exciting cause of melancholia your great endeavour must be to improve the patient's strength and to nourish. In the low depressed melancholic condition which follows acute rheumatism, the patient's excretions must be constantly watched. Alkalies, or vegetable acids, or both, should for a long period be taken by the patients, and the daily exhibition of these drugs is perhaps as necessary as daily food. The same remark applies to gouty conditions. All melancholic patients must be well and warmly clothed, and preserved as much as possible from the influences of damp, and the vicissitudes of the weather.

I have found Cannabis Indica of considerable use when I have had to treat patients suffering from simple menorrhagia : but you must not be satisfied with a symptom, as menorrhagia, amenorrhœa, or leucorrhœa; whatever be the symptom you must endeavour to discover its cause and direct your treatment accordingly. It is often a question whether melancholic patients should not have change of scene, and the question may be answered on broad principles. In all recent cases the patient requires absolute rest, but if you can remove him from his own home it is best to do so. He must be closely watched, and at first must be kept in bed. When he is recovering or convalescent a change is very beneficial, but this change need not be the conventional trip to the sea-side, and certainly should not be a continental tour. You have to remember that your patients have but little physical strength, and that railway journeys and excitement will exhaust them still more. If you can arrange let your patient go a few miles from home in the first instance; afterwards he may go a further stage, and when he can bear travelling he may go further still. Change and variation of scene is of undoubted use in melancholia, but rest and restoration are the first necessity, and therefore above all things physical exhaustion must by every endeavour be avoided.

Melancholy patients will, if allowed, walk till they drop. You must remember this fact, for I believe it is the cause of permanent insanity in many cases that commenced with a recoverable melancholy.

The stage of recovery is usually marked by an improvement in spirits, but, as a rule, the mind of the patient for a long time is very feeble, and it must be brought face to face with the world, and with society and excitement, by degrees only.

One patient who was for some time under my care described her state to me and told me that she felt so enfeebled in mind that she could hardly bear the ordinary sociability of a homely family circle, and that when it was augmented by one or two strangers the social gathering was unendurable. She was an excellent musician, but when in the least degree excited she would cease to play; the notes, she said, became confused, and caused a very painful impression in her mind. The same result often occurred in conversation, when she would become demented and speechless, the power of thought or of connecting ideas for the time ceasing or altogether failing her. On such occasions I never urged her to make any efforts; gradually the feebleness left her and she got quite well. The mental feebleness after an attack of acute brain disease has its counterpart in physical conditions as represented in the feebleness observed after such a disease as fever. A severe attack of fever reduces the physical frame and leaves it in a state of prostration. A severe attack of mental disease reduces the mental constitution to prostration, but when once the attack is over, the patient may rally, and with gentle nursing may return to average health.

General Paralysis of the Insane.

There is a form of disease usually insidious in its invasion, steadily progressive in the character of its course, and certainly fatal sooner or later in its termination, to which the expression "General Paralysis of the Insane" has been applied; some systematic writers have adapted the term "Paresis," or "General Paresis," to this disease, legal writers have called it "Paralysis of the brain," French writers have used the terms *Manie des grandeurs* (a very bad expression) and *Folie paralytic* to denominate the same disease, and Bayle, who first described the disease, called it *arachnitis chronique*. But it seems to me that there is considerable objection to all the terms that have been applied. Pathologists have not yet settled whether these names have been given to one separate and distinct disease or to many pathologies, neither has it by any means been proved that the malady supposed to be indicated by the names, is essentially a disease of insanity, and however unwise it may be as a matter of rule to alter the expressions of common usage, yet in this instance I should strongly advocate a change of term, as the one in most common use, viz., "General Paralysis of the Insane" is calculated to mislead. But I should be sorry to commit you or myself to any expression which might not bear the test of general application, and as the form or forms of disease which have received the name of "General Paralysis of the Insane" is a member or are members of a large class recognized under the head of "Progressive Paralyses," I propose to consider the class, and to speak particularly of certain cases. We shall find that the subjects of Progressive Paralysis are not all insane, but we must study the class in all its forms, and we shall then perhaps gain some broader views of its relation to insanity generally.

Many observers have written positively as to the distinct isolation of "General Paralysis of the Insane," but considerable doubt remains in the minds of others as to the position which the progressive paralysis and insanity, as seen in asylums, ought to take in our nosological tables.

The disease in its various forms has found numerous students, and among them some of the ablest observers, both in this Country and on the Continent.

Bayle, and Calmiel were perhaps the first to note the various symptoms and appearances, and the subject engaged the attention of Ballarger, Jules Falret, Brierre de Boismond, Rokitansky, and Weld. More lately, Professor Westphal, of Berlin, also Drs. Sankey, Lockhart Clarke, Boyd, and Wilks have written on the subject, but the various writers are far from agreement. They have neither reconciled nor differentiated the forms of progressive paralysis as seen in the general hospital and in the lunatic asylum, neither have they all found the same changes in the tissue of the brains and spinal cords which they have examined after the patients' death. If, therefore, we wished to isolate a single and definite form of paralysis and insanity as a definite disease, under a distinct and determinate name, we should have to seek it in the midst of a chaos. It seems to me that, under these circumstances, we shall learn our lesson best by studying cases without present anxiety as to nosology and classification.

The cases, as you commonly see them in lunatic asylums, present such marked characteristics that if you have seen them, and them only, you feel certain that you have a very definite disease before you, and one about which you could never make a mistake, so different is it from anything else; and yet, when you come into the general hospital, you see a disease so like it, the element of insanity only being wanting, that you are bound at once to ask yourself whether the cases you saw in the asylum are members of a definite and distinct group, or whether they do not belong to the same class you see in the general hospital; the insanity being only the accident of the asylum cases?

I would here call to your remembrance an anatomical fact which it is important to bear in mind in the consideration of Progressive Paralysis, it is that the spinal system does not end in the medulla oblongata, but in the corpora striata and optic thalami, and that most of the nerves which we usually call the cranial nerves, belong really to the spinal system, whilst the grey cortex, as it is called, of the brain, which is brought into relation with the central ganglia by means of the white conducting tissue, or, as it is called, the medullary substance, must be considered as essentially the seat of intellectual operations. It will not then appear strange, speaking in broad generalities, that if we have disease affecting the cortical substance of the brain we should have disorder of the mental faculties, and if we have disease affecting the spinal system, we should have disorder of the sensory or motor functions, and of the organs of prehension, locomotion, and special sense: and considering the relation from contiguity of the spinal and the intellectual portions of the nervous system, we cannot be surprised if we see diseased conditions of the one extending to the other. In both cases the disorder of function is, in a greater or less degree, one of arrest, and may be considered in the light of a paralysis, more or less complete according to circumstances.

You will find cases of "Progressive Paralysis" affecting both the intellectual functions, and the somatic organs presided over by the spinal system, and these cases are claimed by alienists as a special disease, peculiarly their own. You will find a form of disease beginning with a progressive mental paralysis, of which alienists predict progressive paralysis of the body. And you see in the wards of this hospital cases of progressive paralysis of the body, of which alienists would predict progressive paralysis of mind; but your experience will be that, though they are sometimes right, they are so only in a comparatively small percentage of cases, and that the mental malady, when it does occur in the general hospital cases, instead of assuming the particular characters they declare it will assume, viz., "delusions of greatness or exaltation of ideas," is more often to be recognized only in the form of a very simple dementia.

The detailed outline of a case, as found in asylums, will give you the best idea of the disease. A patient, G. H., was under my care at St. Luke's Hospital. He was forty years of age—a man of about the average height. He was very well and proportionately developed, though somewhat fat. His head was round and a little bald, and his forehead was somewhat prominent. His expression of face was placid and genial though somewhat vacant, and, except that he had the physiognomy that might on reference to a standard be considered as indicative of "progressive paralysis as found in lunatic asylums," his appearance was pleasing and not by any means that of insanity.

The patient, shortly before his admission into St. Luke's, set fire to his house and was taken before a police magistrate charged with incendiarism; but he was excited and presented evidence of insanity, and the magistrate adjourned the case. Upon further examination, the fact of the patient's insanity was patent, and in consequence he was ordered to be placed under care and treatment. He was brought to St. Luke's Hospital, under the customary order and certificates. When admitted he was feeble and tottery in his gait; he could walk, but only very slowly and with his legs straddled, and he had great difficulty in retaining his balance. He had also lost power over his hands: he could take hold of, but could not grasp or squeeze the hand extended to him, and he was unable to dress himself. In speaking, his articulation was characteristic and peculiar. Some of his words were uttered with very great precision; the remainder were thick, muffled and indistinct, and all were spoken very slowly. The next feature I must request you to mark particularly: I asked him to protrude his tongue, which he did, but he was unable to keep it steady; it presented so marked a tremulousness that the most casual observer would have been certain to notice it. The tongue itself was soft and flaccid, and presented indentations or impressions of the teeth; the lower lip also was tremulous and unsteady whenever the patient spoke or whenever he protruded his tongue; and there was a marked difference in the relative dilatations of his pupils neither of which responded to the stimulus of light. On entering into conversation with this man, I found he answered my first two or three questions logically and well, but when I asked him in what month of the year he was, he did not know. I then asked him if he were rich, to which he replied in the affirmative, and stated that he was worth millions of money, and that he would make me a present of a million. The most curious part of the case was, that on being served with a notice of a Commission " De Lunatico Inquirendo," he appeared very pleased, and when questioned by the solicitor who had charge of the case he gave a correct statement of his property which amounted in annual value to about £800. At the inquiry, he told the Master that everybody was very kind to him, and that he intended to give one million of money to me and another to the Institution. He had forgotten the circumstance of setting fire to his house, and from being a little excited at the Commission, he became depressed and shed tears at seeing some of his relations whom he had not met for a considerable time before. Within ten days of the holding of the Commission I was summoned one morning to his bed-side, and found him insensible as though in an epileptiform seizure, and from this he never rallied. He ceased to breathe within ten minutes.

The post-mortem examination presented all the abdominal and thoracic viscera as healthy; but the cranial contents were striking. The dura mater was very adherent to the skull, and the arachnoid presented a greasy appearance. The brain was remarkably tough and hard, but it was apparently fatty and evidently atrophic. The lateral ventricles contained an ounce to an ounce and a half of fluid. The toughness and firmness of the brain was more marked than in any other instance of " progressive paralysis" I have ever examined. The white matter of the hemispheres especially was tough and the grey matter was thinned to a very remarkable degree. The cord was wasted, and in its coarse change bore a resemblance to the brain. Portions of both brain and cord were set aside for microscopical examination, but, I regret to say, lost or mislaid with a number of other specimens at the time of my leaving St. Luke's.

The points, however, upon which I wish to lay stress at this stage of our consideration are

1st. The paralysis, which was progressive and, judging from the scanty information I was able to obtain, began in the lower limbs. These were more affected than the upper, upon his admission; and the progress of the paralysis apparently was from the feet upwards.

2nd. The tremulous tongue and the thickened speech, or glosso-labial paralysis, and the unequal dilatation of the pupils.

3rd. The loss of memory.

4th. The exaltation of ideas—the form of mental aberration which alienists claim as the peculiar characteristic of the disease which they call "General Paralysis of the Insane."

You will not go into any large asylum without seeing several such cases, and you will generally be told that it is a most distinct and definite form of insanity. In fact, it has been stated that there are only two forms of insanity. The one in which the patients' ideas are warped and askew, they having inherited the defect as another person inherits a squint, but never become paralytic. The other, in which the patients have disorded intellects from acquired conditions which paralyse both body and mind. I, however, believe that such teaching is erroneous. It has certainly not been shown by any careful observation that the ordinary lunatic who spends his life in an asylum, never becomes the subject of progressive paralysis, and it is very certain that very many cases of progressive paralysis bearing the general characteristics of the asylum cases, occur without the patient's giving any evidence of insanity.

LECTURE VIII.

Progressive Paralysis and Insanity.

Alienist's definition of General Paralysis—Dementia Paralytica—Progressive Paralysis—Exaltation of Ideas—General Paralysis without delusions—Early Symptoms—Duration and Progress—Varieties of General Paralysis—Muscular Paralysis Indefinite—Epileptiform Attacks—Pulse—Paralysis of Cranial Nerves—Sphincters—Age of General Paralytics—Progressive Paralysis in Women—Pathology—Brain—Cord—Sclerosis of Nervous Tissue—Progressive Locomotor Ataxy—Atrophy of Nerve-Tissue—Sympathetic Ganglia—Causes—Prognosis—Treatment.

The alienists define general paralysis as "a peculiar mental disorder, with failure of muscular power first noticeable in a changed expression of face and altered mode of speech—a disease of a definite and short duration associated with morbid changes of the brain, cord and membranes;" and their opinions differ very widely as to the nature of these changes. But there is a class of cases which has been described by Dr. Wilks, and called by him "Dementia Paralytica," in which there is failure of bodily and mental power, and in which morbid changes similar to those described by the "alienists" occur, and we may therefore fairly ask what is the relation between the two classes of cases? Some of the cases belonging to the class described by Dr. Wilks, have definite histories, as blows, or shocks from accident or railway collisions. More often, however, the histories are vague and obscure though the symptoms may be characteristic enough; but it is the same with the cases of the alienists. To the same class as Dr. Wilks' "Dementia Paralytica," belong cases of chronic delirium tremens from alcoholism; in fact, the cases described by Dr. Wilks and those described by alienists appear to belong to one and the same class. They are, as Dr. Wilks says, all alike in their fully developed and characteristic stage, feeble, slobbering, helpless, forgetful or imbecile, passing their water and their motions involuntarily, and requiring to be lifted in and out of their bed or chair, and requiring help to get themselves fed.

In all the cases the mental powers become enfeebled, and the memory in particular becomes impaired; but enfeeblement of mind and forgetfulness must not be considered as insanity, otherwise you would brand with a cruel stigma, a large majority who attain to an honoured old age, and a large number who illustrate the effect of the wear and tear of the high pressure living of the present day, though they cannot under any circumstances be considered as mad. In all the cases, whether the "Dementia Paralytica" of Dr. Wilks or the progressive paralysis of the general hospitals, or the general paralysis of the lunatic asylums, there is a visible alteration in the structure of the brain and cord, and this alteration is wasting. Whatever be the exact association, the relation between the symptoms which obtain in the patient, and the pathological change found after death, is one more or less direct of effect to cause—the symptoms which result being the effect of wasting of the cord and brain, definite evidence of which we always find if we look for it after death.

In this Hospital you have ample opportunities of examining facts as displayed in the wards and on the post-mortem table, you have examined numerous cases in Peckham House, and you have seen from time to time, the pathological specimens which I have collected from various sources. You must place the facts you observe in the wards and post-mortem room of this hospital, side by side with those which you learn in the wards and the dead house of the lunatic asylum, and you must weigh and balance and compare the values of both together. In this general hospital all classes of cases are examined; and cases of death from progressive paralysis, both with and without mental symptoms, make their appearance on the post-mortem table, so that here the opportunities for examination are wider and more extended than in the lunatic asylum where the cases are necessarily exclusive and limited to the subjects of insanity.

The visible alteration of structure in the brains and spinal cords of the cases examined by Dr. Wilks, and recorded in the Journal of Mental Science, Oct. 1864, was shrivelling, and this fact, supplemented by the fact that diminished nerve power, exemplified in both bodily and mental failure, was found, on enquiry, to have been *invariably* exhibited in all the patients in whom Dr. Wilks accidentally found this shrivelled brain, constitutes a valuable testament of unbiassed observation, isolating the symptoms we have under consideration by definite relations of cause and effect.

Dr. Wilks' observations bear witness in support of my proposition, that the term "General Paralysis" must extend to a much wider area than the alienist would allow, for Dr. Wilks' cases were not by any means cases for the alienist. You will find some valuable papers by Dr. Wilks on "General Paralysis" and "On some cases of nervous diseases" in the Guy's Reports for 1871 and 1872, which you will do well carefully to study.

The symptoms of progressive paralysis, as exampled by inequality in the contraction of the pupils, tottery gait, tremulous tongue and lips, hesitation of speech, blank expression of face, loss of hearing, forgetfulness, inattention, and inability to concentrate the mind on any subject, certainly indicates that the whole nervous system, including the surface of the brain, the central ganglia and the cord are affected. I do not say that such symptoms always announce that a permanent or organic change has taken place, for the symptoms may be temporary and recoverable, as seen in the cases which sometimes result after concussions from railway or other such accidents. The symptoms, however, when they persist, indicate that organic and permanent change has commenced; and in due time a post-mortem inspection will certainly show the changes I have mentioned, viz., wasting of the brain and cord. And a microscopical examination will in all probability demonstrate the contorted vessels described by Drs. Lockhart Clarke, and Sankey, as also by Rokitansky, and Wedl, which I shall have to speak of and show you by and by. Perhaps thickening of the membranes will be found, but the change found in the membranes often seems to depend on the duration of the case.

The failure of nerve power in many cases goes on to mental feebleness, but this can hardly in the strict sense be termed insanity. Examine such patients and you will find that though they are unable to enter into general conversation, they can answer questions regarding their family and their affairs reasonably and rationally, and they will not talk rambling nonsense, nor show signs of exaltation. They have few desires and few complaints, and if they express themselves as feeling well, it is because they do not feel ill.

On the other hand, insanity may accompany progressive paralysis, may precede it, or may be super-induced upon it.

As found in asylums, progressive paralysis is frequently, though not necessarily, attended with a peculiar delusion, which has been, and by many is considered as pathognomonic of the disease. It is described by the French as *Manie des Grandeur*, and by English writers as "Exaltation of Ideas," or "Delusion of Greatness." But there is a very full and perfect explanation of this peculiar delusion, as seen in association with progressive paralysis, when the fact that it arises out of forgetfulness is considered. But exaltation of ideas often occurs without any paralysis. The exaltation of ideas, or the delusion of greatness, is very well characterised by Professor Westphal, as a silly delirium, the "specific" character of which, as indicating a peculiar form of disease, he refutes, and from his observations he declares that a mixture of stupidity and depression is quite as frequent in the subjects of *General Paralysis* as the delirium of greatness. He ascribes the formation of the absurd and contradictory ideas expressed by these patients to forgetfulness, and to an impossibility in connection of ideas, an impossibility to be comprehended only upon comparison, the patients being incapable of even that amount of connection which obtains in the imperfect logic of an ordinary maniac.

I could hardly find you a better example than that of a patient whom you have already seen at Peckham House, who told you he was going to fill up the sea, but mingled together in his description of his proposed procedure, a tight-rope journey to Australia, pig's food, building materials, clothing, agricultural produce, money and all sorts of incongruities, in quantities of hundreds, and thousands, and millions, without connection or logic, or remembrance in one expression of that which he spoke in the preceding sentence. This man is markedly paralytic. But in another ward you have seen a case to which I have more than once directed your attention, viz., that of a woman who considers herself to be the Deity, the Queen, and various other royal personages, and who fancies that she has power to move the sun. She has undoubted Manie des Grandeur, but she is not in the least degree paralytic. As one of the best tests she is able to play the pianoforte with great execution, a power she had before she became insane, and which she has never lost.

In reference to music as a test, I have in my mind the case of a musical composer of celebrity who became the subject of progressive paralysis, and subsequently of dementia. Shortly after the commencement of his illness, he was altogether unable to play any instrument. One day I placed in his hands a violin, an instrument he at one time played perfectly; he tried to make music, but he was unable to produce two consecutive notes, and what is more, on hearing the violin played by another person, he endeavoured to sing, but was completely unsuccessful in his attempt. He never exhibited any exaltation from first to last.

With the loss of music you see agraphia, or the loss of hand-writing; the patient first forgets how to spell, and then becomes unable to form the letters. The symptom is not peculiar to progressive forms of paralysis, it may occur after cerebral hæmorrhage.

In illustration of the cases which you see in the general Hospital, some of you may remember a man who was in Stephen Ward, No. 44, about this time last year, under the care of Dr. Wilks. When I first saw the patient he was hardly able to move. The paralysis, I understand, commenced in his feet, and gradually crept upwards, destroying on its way the control over the sphincters. The patient could take hold of my hand, but could not grasp it firmly, his tongue was tremulous and indented, his speech was thick, his words were drawled and prolonged, and his memory was defective; nevertheless, he was able to detail with considerable precision his former occupation and engagements. He had been a coachman, and he told me that the symptoms commenced after getting very wet, and that they had continuously progressed for about three years. His physique and his physiognomy much resembled the case I first described. His pupils were unequally dilated, and did not respond to the stimulus of light; his expression was blank and emotionless; he never attempted to move from his position on his back; he exhibited neither mental exaltation nor depression; and he constantly expressed himself as feeling better or well. With the exception of a little boorishness and forgetfulness he presented no remarkable features of mind whatever, and certainly no evidence of insanity. But there were some circumstances connected with his forgetfulness that were remarkable; for instance, he knew how long he had been in the hospital, and the date upon which he was admitted, he also knew the day of the week, but he did not know either the month or the date of the day on which I spoke to him, and said he believed that it was September (it was January), and, when I asked him to enumerate the months of the year, I found that he was unable to do so correctly, but lost their sequence between August and December.

The patient's condition, however, varied considerably from time to time, and it was reported, that on one or two occasions he had an epileptiform seizure, also that once or twice in the night he had some hallucinations of sight. It is said that he imagined he saw fish coming through the ceiling, but whenever I saw him he was calm and free from obfuscation, and I learned that the mental disturbance, which he was said from time to time to have exhibited, was very transient. In all probability, it followed an epileptiform attack. Epilepsy occurring in any patient is quite enough to account for transient mental disturbance.

To say, because a person suffers from progressive paralysis, tremulous tongue, thick speech, and loss of memory, and has the physiognomy of a disease often seen in an asylum, that he is of unsound mind, or that he is suffering from a disease of which mental aberration is an essential part, is to express a cruel, and I think unwarrantable, opinion, and yet I have no doubt but that you would have found some alienists so impressed with the belief of the interdependence of insanity and general paralysis, that they would have signed certificates of insanity against the patient whose case I have just detailed.

The alienist might defend his position by the assertion that if the patient were not giving expression to delusions now, he would at a later stage of the malady, and that he might at any moment break out into extreme maniacal violence, and that already he was unable to take care of himself or of his affairs. This position might be held to be right from the common point of view, but after all I have said, I need hardly now point out that such an assertion has not a good foundation in fact.

The man may not be able to take care of himself, but he is not necessarily mad, nor, indeed, is a lunatic asylum always the best place for him.

There are numerous cases of paralysis general to the body, but uncomplicated with insanity, on record; and I lately saw one in which the patient could hardly speak from the paralysis of the tongue, but all his answers to queries were perfectly intelligent, and I was told that from first to last he had never shown a symptom of mental aberration.

You will find many such cases described in text-books on Medicine under the head of Paralysis. You will find, too, that they usually bear all the features of the cases you see in asylums, the insanity only being wanting; and I think you will be led to hesitate before you draw fixed and limited definition of this disease.

I have dwelt upon this part of the subject because the medico-legal aspect of these cases is of the utmost importance. The question of capacity to make a will or sign a deed has already been discussed many times in our civil courts, and the leaning of expert witnesses has been in favour of the view, that paralysis general to the body is a definite pathology the subjects of which are necessarily of unsound mind.

But I would hold that, if a will should be set aside because the testator was the subject of progressive paralysis, with tremulous tongue and lips and imperfection of memory, or even with slight fatuity, but without further and conclusive evidence of insanity, the judgment was unjust and cruel, and based on insufficient knowledge, and not upon scientific facts.

Progressive paralysis dependent upon central disease does not necessarily begin in the head. The symptoms very frequently first show themselves in the lower limbs ; they are then generally pronounced in the tongue, and show themselves more or less in upper extremities, and after that any or all the cranial nerves may be involved, whilst the mental excitation or depression, and the attendant phenomena of suspension of the power of reasoning or of thought, or the appearance of delusions, do not occur till afterwards ; even loss of memory, the most simple form of mental impairment, may not occur, although the paralysis may be considerable ; in such cases it is certain that the change began in the cord and passed upward.

On the other hand, it is quite certain that in some cases the disease does begin in the head, and in that part of the brain in which the mental powers reside.

The early symptoms of the disease vary. As we have seen, they are of very rapid development, and sometimes, when complicated with insanity, they present very formidable characteristics. There is often a sort of *folie circulaire*. In the first instance a stage of moderate excitement followed by one of depression, which is again followed by excitement more marked and severe than that which occurred before. Sometimes in the excited stage the patients are noisy, destructive, and violent ; sometimes they become the subjects of pyromania, as in the first case I spoke of; sometimes they exhibit the homicidal impulse in a remarkable manner, and the exaltation of ideas, upon which so much has been insisted, may appear as a prominent feature in many of the excited cases. I have not any exact statistics on the frequency of excitement in progressive paralysis accompanied by insanity, but it seems that exaltation of ideas is not more frequent than in thirty per cent. of the cases, such, at least, is the computation of Professor Westphal.

The duration of progressive paralysis varies with its forms. Systematic writers on insanity give to insane cases an average of two years. It is hardly possible, however, to lay down an absolute rule, for although most cases of paralysis and insanity run a course of about two years, some run a course of a few weeks only, and some few others drag on an existence for six or seven or even twelve or more years.

As an example of a rapid case, I may mention a patient who was under my care, and who manifested maniacal symptoms suddenly. He was a clergyman, who laboured to excess in his parish, and expended the greater part of his income among the poor; he deprived himself of nourishment and support to provide funds for his supposed charity, and he avowed that he acted thus from conscientious motive.

At the time that he came under my care he declared himself to be possessed of millions of money with which he intended to build millions of churches.

At the same time he became wet and dirty, erect in his gait but unsteady in the lower limbs, tremulous in his speech and tongue and lips. He was sleepless, violent, destructive; rapidly became feverish and weak, so weak in fact that after two or three days his depression was alarming to the utmost degree, his pulse fell below fifty, and for some days he was kept alive with stimulants: over a bottle of brandy per day mixed with milk, cream or beef tea being administered to him. After the first week of treatment he seemed to rally, in fact he improved and apparently regained strength, so that he was able on two occasions to be up and to go into the garden. He was able then to take solid food, and a moderated quantity of stimulant, but the seeming improvement was very fleeting, within a fortnight he was almost, but not quite completely paraplegic, and within three weeks of the invasion of the attack he died during an epileptiform seizure.

In this case, which was a genuine case for an alienist, most of the leading characteristics were present, and I ask you to note them. The tongue was tremulous and the speech was thickened, the pupils were consequently dilated, ideas were exalted, and the paralysis was progressive, variable, and incomplete; the excitement alternated with depression, the pulse fell to a remarkable degree, and he sustained a number of epileptiform seizures which as we have seen is a constant, if not the most constant, symptom of the disease.

On the morning of this patient's death, when I

offered him some nourishment, he took it, and then declaimed in a loud voice, " I command you to leave me;" the next moment he again commenced to declaim, waiving his hands as though preaching, but his speech was so thick that it was impossible to make out anything he said. He had two epileptiform attacks during the night preceding his death, and a third in the morning shortly before he died and from which he never recovered. I was not able to obtain an inspection, but during his life there was not, from physical examination, any evidence of abnormal viscera in either the thorax or abdomen. He might have had granular kidneys, the existence of which there was no evidence to show; his urine was not albuminous. But though fatal eclampsia from uremic poisoning often accompanies granular kidneys there is no evidence to show that the same cause ever gives rise to the general paralysis of the insane as regarded by alienists, although a wasted brain often goes with granular kidneys. In the case we have just considered there is no reason to imagine that the pathology was of any organs other than the brain and cord. Possibly the pathological change had been slowly, noislessly and stealthily progressing for a long time, and then some accidental circumstance, perhaps an epileptic seizure had lighted up the fire of the acute attack, from which he never recovered.

In illustration of a very slow case, and one which contrasts in point of time with the last, I may mention that of a gentleman who was under my charge in a private asylum. He had been paralytic for five years. The patient became very gradually affected but all through exhibited well-marked mental symptoms, and when at last he became so feeble that he could not move, he would mutter unintelligible sentences in which the expressions "ten-thousand, twenty-thousand," and "fifty-thousand" were all that could be distinguished; he exhibited, in a remarkable degree, alternate stages of excitation and depression, even when the progressive disease had advanced almost to its climax. He would sometimes burst into a laugh, or into a cry, in the middle of his chatter, and without any apparent cause become intensely excited. At the time when I gave up the charge of the asylum the patient was living, but he was so enfeebled that it was evident his end was near. During the five years he had spent in the asylum his general health, as it was reported, might, except for the progressive paralysis, have been considered good; his appetite was good, he usually slept well, and his secretory and excretory organs never exhibited any marked disturbance, and except for the possible pain of mental exaltation and depression, he lived on in his happy delusion, revelling in his imaginary millions, apparently perfectly contented.

The happiness of the delusion, or delirium of greatness is well worthy of remark; when it appears the patients believe themselves to be possessed of vast riches, and what is more extraordinary still, is, that they often believe the asylum they are in to be a palace, their drink nectar, and their food fit for the gods. They will remark upon everything as beautiful; to them everything is *couleur de rose*, even though they be pauper patients in a County Asylum. One patient used frequently to express his delight to me; even the physic he took from time to time was, in his belief, beautiful, the effect of it, he always said, was beautiful, and so were the baths into which he was placed when his dose had told its tale.

A female patient at one time in the City of London Asylum furnished so excellent an example of the vacillation between excitement and depression, that I am tempted to record it. The case was incipient and commenced with a violent and destructive mania, which was followed by depression, melancholy, dejection, refusal of food, and wet and dirty habits. These passed away and a second stage of excitement succeeded, which in time gave place to a second period of depression; this was again followed by excitement and happy delusion, and the patient, believing herself to be a queen, would dance expression to an exuberant outburst of joy. From the first her tongue had been markedly tremulous, and her lower lip also trembled visibly whenever she attempted to speak; but, except under the stimulus of the periodical excitement, I rarely heard her speak at all.

As the mental symptoms varied in her case, so also did the bodily, and though slowly and gradually, it was surely and steadily progressive, and advancing by unmeasured steps towards its certain end.

EARLY STAGES OF EXCITEMENT.

A passing word must be said on the character of the excitement often displayed, by the subject of combined progressive paralysis and insanity, in the early stages; it is sometimes excessive, the cases in fact present the worst forms of mania—excepting only the subjects of epileptic insanity, of whom the so-called demonomaniacs, as we have seen, are, perhaps, the most violent and formidable ever met with, either in or out of an asylum—and at the commencement a considerable amount of method, cunning, and determination, may accompany their violence. I have known the subject of incipient progressive paralysis profess the deepest affection for those around, and act his affection so well as to disarm fear; at the same time he would express his determination to kill those to whom he appeared affectionate, and you will often find that such patients are on the alert to accomplish their purpose, and will seize an opportunity suddenly and when least expected.

W. G., a patient admitted into St. Luke's, with a large unhealthy slough on his foot, exhibited symptoms of incipient progressive paralysis, and before he had been in the Hospital many days became so restless that it was a matter of the greatest difficulty to induce him to keep in bed; if left for a moment he would get up and stand behind the door, and I ordered that one of two attendants should be constantly with him throughout the day, and that he was never to be left.

The order was obeyed, but the patient's restlessness increased, and keeping him in bed was a matter of difficulty. Three or four nights afterwards whilst paying a midnight visit to the wards, the night watches called my attention to a bruise which the patient exhibited on his thigh, and which the sufferer told me the two attendants who had charge of him during the day had inflicted with the chain on which they carried their keys, and he described the manner in which the injury had been inflicted. The next morning I sent for the day attendants and accused them; they however, denied the charge and asked for a full investigation by the Committee, a course I assented to; nevertheless, before the Committee met they absconded, and so succeeded in escaping the prosecution and punishment which they richly deserved. But to return: the patient soon forgot the appearance of the men who had inflicted the injury, and a few days afterwards accused another man with having caused it, and declared he would murder him if he got the chance.

So much for the patient's memory. After a short time I found that he began to sleep better at night time, and, as special watching became unnecessary, I arranged that he should have the ordinary hourly visits of the night watches. One night, however, between the visits of the attendants he happened to be awake and succeeded in reaching a window, having packed up his bedding into a heap on the bedstead, and so formed a platform to elevate himself upon, and by a truly astonishing muscular effort he tore the whole of an iron window frame out of its place, and then broke the ironwork and possessed himself of a bar which he determined to use as a weapon of offence.

The noise made by dislodging the iron window frame from its bed at once attracted the night attendants, and they were opening the door, when a heavy blow against it warned them of danger, and they called for assistance; on looking into the room through an inspection hole in the door, the patient was seen standing in a menacing attitude with his bludgeon raised, and declaring that he would dash out the brains of the first person who entered the room. The attendant, however, whose life the patient had before threatened, threw open the door and closed with the maniac, escaping only by a hair's breadth the blow which was aimed at him, and which, fortunately, struck the door instead of his head. The patient would have made another attempt, but, before sufficient time had elapsed for him to aim a second blow, the bludgeon had been secured.

The following morning he was very much exhausted and made no attempt to get out of bed. But with a coherence of speech that I was astonished at, he told me every detail of his night's exploit. He said " he believed that the attendant whom I have mentioned had inflicted the injury he had sustained, and he determined to punish him by death." He then told me that he had thought "if he could succeed in dislodging the window frame he might make his escape through the hole in the wall, but when he found the hole barred and that he could not dislodge the bars, he bethought himself to break the window frame, and so obtain a weapon to kill his attendant with in the morning; when he found that his noise had attracted a number of attendants to his room he determined to fight, and he would have liked to kill one or all of them, but particularly the one whom he had before threatened to murder." He then charged the same attendant with having committed adultery with his wife, and said that it was his duty on this account to kill the attendant. I asked him if he knew the consequence of murder; he said, "yes" that "murder was punished by hanging, but that he was mad, and that the extreme sentence of the law would not be carried out in his case."

At this time the slough of the foot had increased, and the paralysis of the legs was such that the patient was not able to stand except under the influence of a great excitement like that he had undergone the night before, and he said that "if he could only stand he would get out of bed and kick his attendant out of the room." The luckless servant thus abused was good natured and kind to an unusual degree, and nursed his charge with a tenderness worthy of imitation, and which the patient spoke of and thanked him for, at the same time telling him that it was his duty to take his life.

The patient assured me that he would not injure a hair of my head; but that the assurance was worthless, and that he would gratify his violent impulse upon me or any person, no matter who, whenever the opportunity occurred, was soon proved, for as I was leaving his room he aimed a basin filled with hot beef tea at my head.

He afterwards accused everybody that had anything to do with him of some crime, but more particularly his wife, whom he accused of adultery, though he met her with terms of greatest affection whenever she came to see him. The poor fellow as might be expected became worse and worse, the mania became less and gave place to dementia, but the paralysis increased. I could not, however, follow the case to its termination as the wife took the patient away, and I was never able to hear any more of him.

The subject of progressive paralysis as found in asylums, presents two well-marked varieties of motor affection, the one in which the gait of the patient in the early stages of the disease is erect, with a precise and elevated step, the other in which the patient, from the first, is tottery, his balance unsteady and his step careful. In the first variety, the symtoms have been said to resemble those of progressive locomotor ataxy, and the affection has been considered as ataxy, and described, by Westphal as a distinct form of general paralysis, and called tabic, to distinguish it from the second form, to which he has given the name of paralytic.

The patient at Peckham House, who says he was going to fill up the sea, is a good example of the tabic class. As you observed, he lifts his feet high and brings them down to the ground with considerable force, and if you ask him, he will tell you that the ground feels soft and springy, and if you get him to shut his eyes he will stagger and perhaps fall.

These symptoms may be associated with more or less anæsthesia, but according to Westphal, the patients of this class complain of lancinating and shooting pains in the limbs, particularly in the lower limbs. I have not, however, any note of these pains, except in cases of progressive locomotor ataxy, and they have not been commonly remarked upon by observers of combined progressive paralysis and insanity.

Of the second or paralytic variety the case I detailed to you at the commencement of this subject furnished a remarkable example, the patient could only move slowly, his legs and knees being stiff as though confined in splints, and although he could keep his balance with his eyes shut he could not retain it if his legs were close together, and when he attempted to walk he would keep them straddled or widely apart. You have seen several of such cases at Peckham, and so I need not dwell upon them.

If you examine a number of patients of the class called tabic or ataxic I think you will find that very few examples of genuine locomotor ataxy are present among them. I have seen one case of typical locomotor ataxy become insane, and I had the histories of two more from a most reliable observer, and I have heard of others, but they are rare. The common form of progressive paralysis which is attended with insanity, does not develope the characteristic symptoms of progressive locomotor ataxy, viz. the pain, and the loss of co-ordination and precision; the patients have ataxic symptoms, and exhibit great loss of control over their movements; as a rule they lift their feet high but they bring them to the ground with great precision.

The upper limbs are never as perfectly paralysed as the lower, but in them the loss of both motor and sensory power is often very characteristically exhibited; not only is it indicated in the agraphia and tremulous handwriting already mentioned, but it is shown also in the patients' inability to dress themselves, to button their clothes, or to feed themselves. The arms whilst lying still may not convey any notion of paralysis, but as soon as they attempt any voluntary movement they commence to tremble. You will best observe this if you get a patient to try to feed himself; you will see that he can scarcely bring the fork or the spoon to his mouth, whilst by a series of tremblings and jerks he will spill or drop the contents.

The trembling, however, is not due to the same pathological condition of the cord as paralysis agitans. You will be able to compare my sections of cord from paralysis agitans with those of paralysis and insanity, and you will see that they are very different; the paralysis agitans being associated with a very peculiar form of sclerosis, and a remarkable metamorphosis of the cells.

The common opinion that a true muscular paralysis exists, is, I believe, an error. The apparent muscular paralysis is rather from a loss of control than from a loss of power to contract. I was first led to this belief from the fact that numbers of the cells of the cord appear healthy on microscopical examination, and on testing the pulling power of patients who had sufficient sense to respond to a request, I found that the muscular power remained good. You will have the opportunity of making this observation for yourself, at Peckham House, if you have not made it already.

I have now to direct your attention to some characteristic features, most of which, however, are common to all forms of progressive paralysis, whether complicated with insanity or not.

A characteristic of progressive paralysis, particularly that form in which the mind becomes involved, is epilepsy; almost all, if not all, patients suffering from paralysis and insanity become the subjects of epileptiform seizures. The epilepsy is, no doubt, due to the wasting of the brain, but it is not confined to insane cases of progressive paralysis; sometimes the epileptiform attack is the first announcement of the disease, at other times the epilepsy does not show itself till the disorder is far advanced.

Dr. Moxon had two excellent cases in the clinical ward in the summer of 1872, both of which illustrated the epileptiform characteristic of the disease remarkably well. In the first case the malady apparently began with epilepsy, which was followed by slight mental disturbance, though he was not insane when I saw him. Nevertheless he had progressive paralysis, and well illustrated the tottery gait. He also had the irregular pupil, and the glosso-labial paralysis I have already referred to, as well as the characteristic agraphia, he was unable to write even his name. The other case likewise commenced with epilepsy as a first symptom, and illustrated the progressive paralytic symptoms remarkably well. He furnished well-marked evidence of mental disturbance, in hallucination of sight and sound—seeing wheels of fire before his eyes, and hearing voices; but such hallucinations were not delusions, for he was able to correct the errors, and in speaking of them he described them as fancies; and at the time I saw the man in the hospital I could not say that he was mad. He shortly, however, became very obstinate and unmanageable, and at night time was delirious and noisy, and it was found necessary to remove him. I learned afterwards that he was placed in an asylum and, as I anticipated, soon died.

At the time I saw this patient, I expressed an opinion that the case would terminate rapidly, and, *inter alia*, I was led to this opinion from the man's pulse which was characteristic. I have observed that in rapidly fatal cases the pulse falls in an extraordinary manner. In Dr. Moxon's case the pulse was sometimes below 60 (in two cases you have observed in Peckham House, the pulse was 64 in one and 68 in the other). I had two cases under my care, one in St. Luke's, the other in a private asylum, in which the pulse was below 50, and both died within a few weeks of the first appearance of recognisable symptoms.

A characteristic upon which I would for a moment dwell is the alternation and variation of the symptoms. Both the mental and the bodily symptoms vary, and to such a degree that patients who you at one time believe to be dying, will so far recover as to appear well. But they always relapse and the re-appearance of the symptoms with renewed force after a time will go to show that the progress of the disease was, in reality, not arrested. The mental symptoms, in particular, when they exist, exhibit very marked stages of alternation. Sometimes very marked excitement, at other times very marked depression succeed one another, whilst often there are intermediate stages of calmness and placidity during which it is not easy to make out mental paralysis at all. You saw a case of this kind the other day at Peckham,—the man who said that he was perfectly well, and that he had never suffered anything. You might have made out a slight tremulousness in his tongue, and hesitation in his speech, but nothing more, yet he has been markedly paralytic, and no doubt will be again.

A very important characteristic is the tremulous tongue and glosso-labial paralysis.

Much stress has been laid by some observers upon the tongue and lip paralyses, and they are very valuable symptoms of progressive paralysis, but they do not belong exclusively to the insane cases, and are by no means pathognomonic of a peculiar mental disease. The irregularity of the pupil so often seen in the insane cases is likewise found to accompany all forms of progressive paralysis, but any of the cranial nerves may become involved as the palsy creeps on. In ordinary cases the nerve of taste probably is soon paralysed, and the pneumogastric can hardly escape though its involvement may be late. Paralysis of the seventh is often early exemplified in facial paralysis and deafness, of the fourth in ptosis, of the sixth and third in disorders of vision from defect of focus and strabismus, whilst amblyopia, or amaurosis occasionally gives clinical evidence of the involvement of the second. Among these symptoms you will observe none that may not be present in the cases which never become insane. I would also call attention to the facts of the paralysis of the sphincter muscles, and the tendency to bed-sores, in which many of the cases clinically present a strong likeness to ordinary myelitis.

Another important characteristic is the patient's age. Progressive paralysis is not a disease of old age or of youth, but of middle life; its average period being between 30 and 40. It is seen also to attack the strong and vigorous, and rarely, if ever, appears in the ordinary subjects of delicate constitution or in patients suffering from diseases of the thoracic or abdominal viscera.

It must also be noticed that in women progressive paralysis with insanity is not common. You have seen some cases at Peckham, and you will find some in every large asylum, but there are very few cases on record in which the disease has appeared in female subjects belonging to the upper walks of life. I have seen one, and only one, characteristic case in a woman of the middle class of life, and all observers have noted that the affection is rare in women except amongst those of the lower classes. It happens, however, that all forms of progressive paralysis are more rare in women than in men. But what determines the selective preference, as it were, for the male sex, has not been satisfactorily explained.

It has been asserted that a cause of progressive paralysis and insanity is excessive sexual excitement, and that women are more fitted for bearing sexual excesses than men, and therefore less liable to progressive paralysis as a result. I think a very different conclusion may be drawn from the observations on venereal excesses. May not the recorded sexual excesses be regarded as the outcome of nervous disorder? If the paralysis seen in lunatic asylums be the result of excess of venery, why do not all people who indulge in sexual excesses become the subjects of insanity and paralysis? Examples of sexual excesses are constantly met with, but whatever be the result of such indulgence it is by no means common to see the subjects of it become insane paralytics.

Pathology.

The pathological anatomy of progressive paralysis, particularly of progressive paralysis combined with insanity, has been studied by many observers, but our knowledge of it is not yet as definite as we may hope to see it. The various appearances have yet to be separated and grouped, and the processes of change explained. Of the contributions to our knowledge of the subject, those by Dr. Lockhart

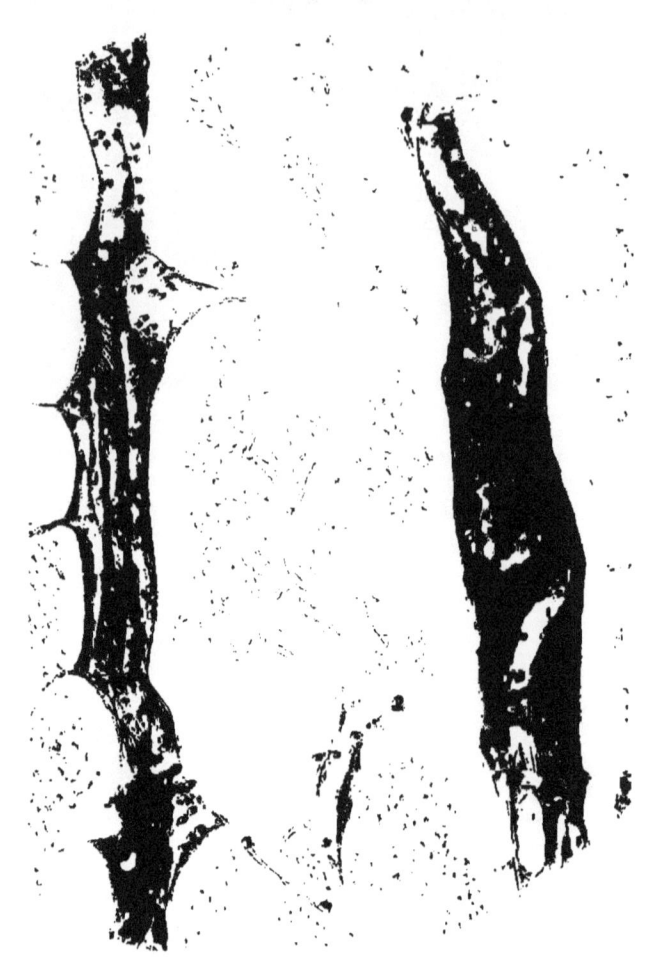

Clarke deservedly take the first place, and I have already mentioned the labours of many others. In this country Drs. Sankey, Salomon, Wilks, and Boyd have furnished the results of their observations, and on the continent, Rokitansky, Wedl, and Westphal, also M. Robin and M. Joire, Drs. Poincari, and Henri Bonnet have furnished records. But many of the various investigators have arrived at different results, and many statements have yet to be verified.

I have myself made numerous observations, and from my specimens which are before you, you will be able to form some judgment for yourselves.

In the early part of this lecture I mentioned that the post-mortem appearances were those of wasting, or atrophy and shrivelling of brain and cord, and formative changes in the membranes. You will see from the portions of brain and cord on the table the shrunken appearance. The brain numbered "15" was from a typical case, it was sent to me by Dr. Crichton Brown, of the Wakefield Asylum, and it is small. On holding thin sections of it up to the light you see numerous slits as it were, and these which were first observed by Wedl, and Rokitansky, were afterwards studied by Drs. Salomon and Sankey, who described tortuous vessels in them. Dr. Lockhart Clarke has since more fully described the appearances, and has shown that the slits are widened perivascular canals, or the perivascular lymph spaces which were demonstrated by His, which have become widened. In these widened perivascular spaces the vessels become twisted and tortuous, sometimes they are so contorted as to be formed into links or knots; around these twisted vessels Dr. Clarke has shown that there is a loose investing sheath of fibrous or areolar tissue, or hyaloid material, and that numerous granules of hæmatozin are deposited within or upon the sheath.

You will see these conditions in all the sections on the table. The specimen marked G. P. C. is one which Dr. Clarke kindly lent to me for the purpose of showing to you, and the drawing in colour is a representation of it. You will remember that I exhibited specimens, showing similar changes, to you in our consideration of melancholia, in which, however, the vessels were not so greatly convoluted, though the spaces were very wide and the hæmatozin was present. The sections marked 21 show the same conditions most distinctly and they curiously enough are from a brain which I received as healthy.

I have already mentioned to you the thinning of the convolutions. (Dr. Herbert Major of the Wakefield Asylum, has invented an ingenious instrument for measuring the depth of convolutions, a description of which you will find in the British Medical Journal, Aug. 24, 1872, and it may become of value to render exact future observations). In this thinning or wasting, changes occur in the cells. Dr. Clarke describes them as consisting of "an increase in the number of the contained pigment granules, which in some instances completely fill the cell. In other instances the cell loses its sharp contour and looks like an irregular heap of particles ready to fall asunder.

On the table are numerous sections in which you can see these granular cells, and you can compare them with the healthy sections. Also on the table, in one of the drawings, there is a fair representation of the morbid cells, which will help you to recognise the unhealthy condition.

The pathology presented by the cord has been read in different lights by its various observers.

Dr. Lockhart Clarke has found it softened, in some instances, and states that he has noticed it of a consistence like cream. In other cases he has found areas of fluid disintegration in the grey substance.

Dr. Boyd, in a paper on the subject in the *Journal of Mental Science*, testified to a condition of softening, as found in some cases he examined. Professor Westphal of Berlin, who discussed the various theories which have been advanced, in a paper ably translated in the *Journal of Mental Science* for July, 1868, and January, 1869 by Dr. James Rutherford, states that he has frequently found grey degeneration of the posterior columns, or of the posterior and lateral columns, and sometimes atrophy of the nerve roots, and some of his cases have appeared to him to bear, in their pathological anatomy, a strong resemblance to that of progressive locomotor ataxy.

Some observers have believed that a general sclerosis of the brain and cord takes place, and state that there is an increase in the areolar tissue, we must therefore consider the various points.

T

I hand round to you two pieces of cord marked 2, and 18, from patients who suffered from progressive paralysis and insanity. You will see that these cords have never been soft, there was no bulging of the nervous substance on cutting them when fresh, as occurred in the case of specimen marked No. 3—the cord of a man who died from fever. When, or how, we are to draw a line between pathological and post-mortem softening I do not know. The evidence of softening presented to the eye, or to the finger, on taking a cord out of the body, is most unsatisfactory, and after the substance has been hardened in chromic acid the microscope does not demonstrate any differences between pathological and post-mortem softening. I have obtained both fibres and cells looking quite normal from cords as soft as cream, and I therefore think that we must not regard softening of the cord as the essential pathological change associated with progressive paralysis and insanity. The idea that sclerosis is the condition of the cord in the form of progressive paralysis which is often associated with insanity is an error. If you will examine the specimen marked No. 2 d you will see, in the dissection, the fine areolar tissue] which holds the bundles of nerve fibres together, but you will not find any excessive quantity of this areolar tissue: the microscopical specimens from the same cord will show you the same fact, and if you will compare the sections from that cord and marked Case No. 2 with the sections marked Case

No. 11a, 2 and 4, sections of cord and brain from a typical case of sclerosis, you will at once be struck with the difference. In the first set, in section Case No. 2 a, you see areolar tissue, but the specimen has been prepared expressly to show it, and it is not to be seen in other sections of the same cord. But in the sections of the other set—Case No. 11, that of sclerosis—you can see very little else than areolar tissue, and the sections had no special preparation.

The specimens marked Case No. 4 a are sections of cord from a case of progressive locomotor ataxy, and Case No. 4 b are sections from the cerebrum and cerebellum of the same case. You will see that they present no sort of resemblance to those marked Case Nos. 2, and 18, and 15, and G. P. C., or the sections of the cords and brains from the typical cases of insanity and paralysis.

Professor Westphal's cases must have been examples of progressive locomotor ataxy, the subjects of which became insane. I have already mentioned that I have seen one such case, and have obtained reliable information of two others. The cases of grey degeneration of the lateral columns, too, which he mentions, may have been general sclerosis associated with insanity. General sclerosis is quite sufficient cause for a paralysis general to the body, and when the change has attacked the brain, as it has done in the sections before you, I see no possible reason why the mind may not become involved also, though of course it does not necessarily do so, unless the cells become affected.

The somatic condition, however, which goes with general sclerosis, though a general paralysis is not the affection commonly seen in asylums.

You will see the anatomical character of sclerosis, or, as it is sometimes called, of grey degeneration, best in the posterior columns of the cord, from a case of locomotor ataxy, the sections are marked Case No. 4, or in the specimens of general sclerosis numbered Case 11. You will see that the change is formative. An areolar tissue having been formed between the nerve tubes and contracted, has dragged upon and dilated the tubes in a similar manner to that by which the tubes of the lung become dilated in cirrhosis of that organ.

The white matter of Schwann is in all probability increased, and you can see in some part of the sections the perfect outline of the tube surrounded by the hypertrophic or increased areolar tissue, and in the centre of the tube you see the axis cylinder. You will observe differences in the change in various parts of the sections, but these differences are merely differences of degree, and illustrate the stages of advance.

The specimens marked Nos. 2, and 18, are from typical cases of progressive paralysis and insanity, as found in asylums, and are very characteristic. The change is principally in the white matter which is shrunken and wasted. You will see that in some parts the axis cylinder remains, but the white matter of Schwann is deficient, so that the nerve tubes are of very small diameter, shrunken together, and arranged very irregularly, in other parts the axis

cylinders are only to be seen irregularly distributed in the tissue, whilst it is impossible to distinguish the outlines of the tubes. The white fibres again are shrunken together, and collected into bundles, leaving vacuoles and spaces in which tortuous and convoluted vessels are to be seen coursing along. In one of the cords—the one numbered 18—the spinal canal is enormously dilated, and, every here and there a little collection of transparent material, which may, or may not, be a pathological formation, is to be seen in the canal. In the grey matter you will notice several collections of similar transparent substance bearing the shape, and occupying the position, of cells, and looking as though they had taken the place of cells that had disappeared. This is the fluid metamorphosis, spoken of by Dr. Clarke and some of the French authors.

The cells of the insane cases marked Nos. 2 and 18, show here and there an increase of pigment, or a granular change, very many of the cells, however, appear quite healthy, and you can trace their caudate prolongations for some distance as they run among the transverse fibres of the grey substance which are in some of the sections remarkably visible.

The condition of the cells, generally, is very unlike that found in the cords of patients who have suffered from progressive muscular atrophy, a specimen of which is marked No. 5, in it you see the cells have dwindled or become attenuated, but have left many of their sheaths intact to represent them.

The wasting of the nerve roots, as described by Westphal, is not very definite, and the ganglia on the posterior roots are certainly sometimes unchanged. Sections No. 18, G, are through some of the ganglia of the posterior roots from the cord marked No. 18, and they appear quite healthy. Upon this point, however, I have only the negative evidence of the sections before you to go, so that I cannot speak very definitely as to the general rule. That the change in both brain and cord is one of wasting, there can be no question, and that it is primarily and essentially wasting I think can hardly be doubted. There is still a wide field for observation and investigation as to the manner in which the change is brought about, and as to the constitution of some of the products of the change, such as the numerous granules and globules which appear on the field of the microscope. I have already tested many sections with various reagents, but have not arrived at definite results, for the globules which look like oil are unaffected by ether or benzole in which I have soaked specimens for hours, and the granules which look like corpora amylacea are unaffected by iodine. Whatever be the exact constitution of the little bodies you see, they, no doubt, are the product of the shrivelling and wasting change.

Whether, at an earlier stage, a chronic inflammation of the cord occurs, has not been shown. Some writers have attempted to prove by the temperature of the patient that such was the case, but the evidence, when closely examined, was hardly

admissible, its value being extenuated by modifying circumstances. There is no doubt, however, that atrophic change occurs, and it is possible that an antecedent inflammatory change may occur, but in the absence of the products of formative inflammation we have no grounds for surmise. There is, however, evidence of formative change in the membranes. The specimens you have before you—Case No. 2—and the drawings I hand round, show you a most remarkable degree of thickening, particularly of the dura mater, and point to a pachymeningitis of a very chronic type.

In the brain, although we find wasting shown by loss of weight and shrinking, (the tendency to vacuum being compensated by fluid effused into the sub-arachnoid space,) yet the volume is not always decreased. Sometimes the brain seems too big for the skull, as in the case I described at the commencement of the subject, the cause of this increase being due, generally, to œdema. Occasionally the brain appears swollen; and sometimes, although it is wasted and shrunken, it is fatty, and, in consequence of the deposit of fat, it is not much altered in volume.

Sometimes traces of an inflammatory process are observed on the surface of the brain, to which the pia mater may be adherent; and, like as I have shown in the membranes of the cord, there is sometimes evidence of a chronic inflammatory process in the membranes of the brain also—they are variously thickened and adherent, the arachnoid becoming partially opaque, and the dura mater sometimes exhibiting hæmatin staining. With these changes are found granulations of the arachnoid, and it is also common to find the dura mater intimately adherent to the skull.

Looking at the pathological anatomy as we have it before us, we can but associate the paralytic symptoms with the atrophy in the cord, and the mental symptoms we must associate with the wasting of the brain; some modification being occasioned perhaps, in particular cases, by the inflammatory affection of the membranes.

It is plain, however, that we may have a wasting affection which may extend through both cord and brain; and, judging from clinical facts, we may conclude that the affection may start in either brain or cord; but there is no evidence to show that the involvement of one is necessarily followed by the involvement of the other.

There is one fact of pathological anatomy in connection with this disease, which has been brought to the surface by Drs. Poincarè and Henri Bonnet, which gives promise of shedding some light upon the starting point of the affection, and may lead us by and by not only to a better recognition of the various forms of progressive paralysis and nervous diseases, but also to better differentiations, and more exact pathological conditions.

These observers have studied the condition of the vaso-motor nerves, and I extract from the translation of their paper as published in the *Journal of Mental Science* for July 1869. They say " In general paralysis the cells of the whole chain of the great sympathetic are covered with brown pigment to a degree much more intense than in other subjects, from whatever disease they may have suffered. In the ganglia of the cervical region, and often in the ganglia of the thoracic, there is evidently a substitution of cellular tissue and of adipose cells for the nerve cells, which last are comparatively rare.

Everything leads us to think that this is the anatomical starting point of the affection, and that the alterations of the encephalon are the mere consequences of the disorder which this sclerosis, by a paralytic action of the cervical ganglia, produces in the cerebral circulation. There is always a very marked pigmentation of the spinal ganglia and of those which are attached to the cranial nerves. The adipose cells, which are substituted for the nerve cells in the ganglia of the great sympathetic, often exhibit a depth of colour which may even be quite black."

This is a valuable record, but I cannot pass it by without some criticism. The authors have stated rather positively that the change is one of sclerosis, and this point is open to doubt. Fat cells, pigment cells, and areolar tissue, they undoubtedly found, but whether they found an increase of areolar tissue, which alone can constitute sclerosis, is, I think, open to question.

There is yet one more pathological point which I must call to your remembrance and remark upon in particular, it is the fact that epileptiform seizures almost always, if not always, sooner or later accompany this disease. As a clinical fact, the epilepsy produces all the curious and fleeting manifestations of complete or incomplete paralysis, as we saw in speaking of epilepsy and mania. We may see a limb or a side paralysed, or in some cases we may see the tongue attacked, affecting the speech; and if the patients die in a fit their brains present evidences of anæmia, serous effusion, or hyperæmia, according to the stage in which the fit existed at the moment of death. It is, however, consistent with clinical observation, as I have already pointed out, that progressive paralysis may be ushered in by a fit; the symptoms both bodily and mental may be aggravated by a fit, or the patient may partially, or for a time, apparently recover after a fit; and since a person may have a fit in the night which nobody knows of, it is possible that many instances of brain degeneration and general paralysis are ushered in with epileptic attacks which cannot be treated. The relation between progressive paralysis and epilepsy is a field for extended study well worthy of cultivation. We want more facts as to the progress of the cases which have been grouped together as paraplegia, many of which I believe have wasted brains, and are epileptic. I have under my care a patient who has suffered from progressive locomotor ataxy for three years, without presenting any mental symptoms, but in whom, lately, nocturnal epileptiform attacks have become manifest, and with them loss of memory. Had we more records I have no doubt we should find wasted brain and epilepsy in connection with all forms of progressive paralysis much more frequently.

Etiology.

Beyond the possible though obscure origin of some of the cases in induced epilepsy, the causes of general paralysis are unknown, and no statistics that have yet been collected have shed any clear light upon the etiology. Excesses of all kinds, and physical shock are among the most prominent of the assigned causes: alcohol, excessive venery, onanism, and syphilis, all get the credit of setting the morbid change agoing, but facts to prove the assertion are wanting. Some of the patients it is true indulge in masturbation, and some, in the early stages, indulge in excess of venery, but these, as I have before remarked, are evidences, rather than causes of the disease. Sun-stroke, which appears in the history of some cases, must not be omitted from among the possible causes. After sun-stroke the patients sometimes suffer from symptoms of progressive paralysis affecting both body and mind, and then get well, but the sequelæ of sun-stroke are too variable to draw any general conclusion from.

Sometimes the cases are referable with a questionable presumption to accident, such as blows or shocks, and it seems certain that over-work or over-stimulation of all kinds may be seen commonly in connection with it; but from the unravelling of the mystery of the subtle influence which sets up the pathological process, even though it start in the sympathetic, we are as far off as ever.

The disease appears in people possessed of high intellectual capacity, as well as those of but mediocre ability, though it seems to spare the subjects of very small mental power. Some of our greatest minds have been shattered with this disease, and it appears in families of all grades and classes, from the nobleman to the pauper.

I have noticed, in a considerable number of the cases I have had the opportunity of observing, that the persons who have become affected with this disease have not any issue, this is by no means a rule, but as I mentioned, it happened in a considerable majority of the cases I have seen. Some further observation on this point may help us to a solution of the question whether the disease is hereditary or whether it is acquired by the individual. If it be shown that a large majority of the subjects of progressive paralysis are incapable of procreation, it points to the conclusion that the disease is one which has a tendency to die out, and therefore probably has its origin in the individual, and it seems very probable, although we have but scanty facts to judge from, that though the paralytic may procreate beings who become similarly affected, yet that the morbid species cannot be continued into a third or fourth generation. The prognosis of general paralysis is always unfavourable. I have heard of recovery, but have never seen it, and the cases, if they exist, are very rare. Many cases, however, do improve, and get nearly well, or for a time appear quite well. You must, however, place but little reliance in the permanence of such a seeming good result, unless you have evidence that the symptoms were only those of a temporary condition. Your patient may have appeared quite well for two or three years, after which he may have an epileptiform seizure, when all the symptoms return in their full violence, and the disorder rapidly runs on to its end. You will have sometimes to consider apparent recovery very seriously, for patients will ask you to make affidavits, with a view to a petition of supersedeas, praying that a Chancery decree may be set aside. I recommend you to exercise great caution in making such an affidavit, for the probability of permanent recovery is almost nil, and the probability of relapse is exalted to the maximum.

The patients often die in a fit, sometimes they die from exhaustion, and sometimes from the complicating results of defective innervation, such as bed-sores, or hypostatic pneumonia. The condition of the patients during the progress of the disorder is very variable, sometimes they grow fat, sometimes they become emaciated, and sometimes they alternate between the two stages, at best, however, they are feeble and weakly, and susceptible of the impressions of very slight external influences.

Treatment.

The treatment of progressive paralysis can only be palliative and hygienic, for the tendency of all forms of the malady is progressively downwards to its termination.

It is however a matter of encouragement and importance to know that palliation is sometimes possible, and in the early stages we should use, with strenuous efforts, all the means at our disposal.

The first great therapeutic agent is rest; and if the patient can at once be relieved from all labour, both mental and physical, he will be placed in the best position for recovery.

The next indication is support; the patients should be well fed, and may have a moderate amount of stimulant; but the effort must be directed towards general healthy nutrition and not to fattening, and it must be remembered that the patients often exhibit a disposition to become fat, even to the involvement of the tissues, and in particular the bones.

The success of the use of medicines is very doubtful: iodide and bromide of potassium appear, sometimes to have been used with benefit; perhaps when the symptoms have followed syphilis these drugs may prove useful. The calabar bean has been prescribed with some success in the early stages, but it is only a sedative after all, and does little but relieve urgent symptoms. Strychnia has been used but it is too stimulating; and blisters over the spine have been used; Dr. Boyd, having regard to the inflammatory conditions found in the cord and its membranes, applied them to cases in the Somerset Asylum with some success, but bearing in mind the liability of the skin of paralytics to slough, the greatest care must be exercised in the use of irritants. From my own observation I am inclined to look upon the success of the blister treatment as very doubtful. The continuous current of galvanism may be applied in the course of the cord to early cases, and it may for a time restore a very considerable amount of tone to the tissue; faradisation, however, must be avoided, it never appears to do good and it may do positive harm. Douche baths and shower baths will often tend greatly to relieve, and give tone to the cord and comfort to the patient, but you must always remember that there is a point at which benefit from all applications ceases, and that, when that point is reached, it is wisest and best to leave remedies alone altogether.

Maniacal symptoms and symptoms of depression must be treated according to the methods I spoke of under the heads of treatment of mania and melancholia. It must, however, be borne in mind, that the patients suffering from paralysis and insanity require even more care than patients suffering from insanity alone. They are helpless, and they may fall and bruise themselves, and as their bones are often fatty and brittle, their ribs may be broken by a fall or by rough handling, whilst the anæsthesia is such that they may be considerably injured without complaining. You must remember, and I would again lay stress upon the fact, that progressive paralysis is a disease the characteristic of which is failure of nerve power, and calls for gentle and good nursing in proportion to the nerve failure, for good and judicious nursing may be the means of considerably prolonging the patient's life.

In conclusion I would repeat that in progressive paralysis, the loss of motion and sensation is always incomplete; it may, or may not, give expression to failure of mental power, but if it does, the defect may range from a slight loss of memory, through some conditions of forgetfulness, fatuity, mania, and melancholia, to the most absolute and the most profound dementia.

LECTURE IX.

DEMENTIA.

Dementia Defined—Comparison between Dementia, Imbecibility and Idiocy—Acute Dementia Illustrated—Pathology of Acute Dementia—Causes of Acute Dementia—Chronic Dementia—Forms—Sequel of Acute Mental Disease—Primary Dementia—Infantile Dementia.

There is a state called dementia which I would define as loss of mind, to which we must devote some of our attention. The subject is important, not only from the variety of its forms and the range of its subjects, but from the prominence of its proportion among the subjects of diseases of mind.

The term dementia is very definite, yet, strange to say, there exists much confusion in its use, it has been erroneously applied to every abnormal mental state, from acute mania to idiocy, but as implied from its derivatives *de* out of, and *mens* the mind, it is applicable only to the cases in which there is loss of mind.

It must be distinguished from madness or perversion of mind, as mania or melancholia, and from amentia, as exampled in the condition of idiocy in which the mind has never been developed. An idiot is not the subject of dementia, but of amentia, because his is not a case of the loss of a mind he once had, but an absence of mind, because he never possessed any. Much confusion indeed exists in the common use of the terms imbecile, idiotic, and demental, but they are definite enough and ought not to be used convertibly. Esquirol was the first to point out the confusion of terms which exists, and he remarked " dementia must not be confounded with imbecility or idiocy. In imbecility, neither the understanding nor the sensibility has been sufficiently developed. He who is in a state of dementia, has lost these faculties to a very considerable degree. The former can neither look backward nor into the future ; the latter has recollections and reminiscences. Imbeciles are remarkable by their conversation and acts, which greatly resemble infancy. The conversation and manners of the insensate bear the impress of their former state."

If we used words with more precision we should have fewer difficulties in studying medicine.

In the subjects before us we must confine the application of the terms, using *Dementia* to express loss of mental faculties which once existed.

Imbecility, defect of mental power owing to the congenital disease.

Idiocy, absence of mental faculties from want of development.

The distinction rests upon pathological grounds, and the answer to the question of healthiness or unhealthiness of tissue often determines to which class the case belongs.

Idiocy is a condition of amentia, or the absence of mind because it was never developed, and it co-exists with undeveloped, though not necessarily unhealthy conditions of brain substance. As a rule the amentia is more or less comparative; a greater or less degree of mind is usually developed in idiots, for those who are absolutely amental commonly die in infancy.

Imbecility properly applies to the subjects of congenital disease of the brain, or to conditions whereby the brain becomes unhealthy whilst the fœtus is in the utero, in consequence of which the perfect development of the mental powers is impossible, and the subjects exhibit a small amount of intelligence, or feebleness, or weakness of mind as it is commonly called. If, however, the case be one of the loss of mental faculties which once existed it is very properly called dementia. The precise limits are not always perfectly definite as the cases sometimes run into one another, but the terms in their generic relations to diseases of the mind are clear, and their use is fully warranted by their expressiveness.

Dementia may be acute or chronic, and there is a third form of the disease to which the adjectival expression "infantile" has been applied, and it is a useful term and sufficiently indicative of its special condition.

The first class, or acute dementia, occurs as a primary disorder. It may, and often does, make its appearance quite suddenly, and sometimes it is irrecoverable.

The second class, or chronic dementia, appears in two distinct varieties. The one, more or less a secondary condition, following either acute dementia or mania or melancholia. The other, a primary disorder, which begins insidiously and progresses hopelessly, and with it are grouped the cases of dementia which are the legacy of old age, of some cases of epilepsy and other deteriorating nervous disorders.

The third class, or infantile dementia, is very frequently confounded with idiocy and imbecility. But it is a distinct condition. It is a true dementia, having its origin in infancy, and its nosological relationship to imbecility and idiocy is highly important.

In illustration of acute dementia, I will detail some cases.

In the month of February, 1872, a youth, æt. about 20, was admitted into this hospital under the care of Dr. Rees. He had previously been in the hospital suffering from morbus cordis and was discharged relieved, but the next day was brought back in a cab by a policeman, who found him in a fainting condition near a public house in the Old Kent Road. He was first taken into the public house where some brandy was given to him, but when spoken to, he did not answer. His mother, who on missing him, enquired for him at the hospital, stated that he had left his home without giving any reason for doing so; and he appears to have wandered on until he was exhausted. When questioned at the time of his admittance he gave no answer, and after being put to bed it was found that he would not feed himself, and that he passed his motions and his urine under him. He continued for days silent and placid without making the slightest effort to help himself. He took food when it was put into his mouth, he sat up in bed when lifted up, or in a chair when placed in it; but he did not make the least attempt to move himself, and unless moved from time to time he never altered his position.

I examined this interesting case one day with Dr. Moxon, under whose care he had been on a former occasion. We found the lad suffering from mitral disease. He was lying in bed, motionless, silent, and placid, he was not insensible, and the calm expression of his face might almost have been considered as intelligent. His eyes moved, and his face conveyed an expression of comprehension when he was spoken to; but he did not give assent either by word or gesture. He would look or stare, but was otherwise unmoved by everything that occurred around him. He did not answer when spoken to; but there was no history or evidence of deaf mutism, and his emotionless demeanour and the profound silence he maintained, were in the utmost degree singular. I was told that he had been galvanised, the effect of which was to make him scream, but not to induce him to speak.

Dr. Moxon asked him to write his name, which he appeared willing to do, and he wrote "George" correctly, but after writing half of his surname he appeared to have forgotten the remainder: he was then shown a watch, and asked what it was and after much coaxing and persuasion he wrote "Gold": he was then shown a drinking mug and asked what that was, and in answer he wrote the word "pot": he was then asked if he could speak, and he wrote the answer "yes": he was then asked if he would speak, and he again wrote the answer "yes"; but no persuasion would induce him to utter a single word. He was then offered a shilling which he looked at but made no effort to take, when it was put into his hand he held it, but without moving his hand, and when asked if he would give it to his nurse to keep for him he made no gesture; when at last the nurse was told to take charge of the shilling for him he allowed it to be taken from his hand without the least concern. The case for some time appeared like aphasia, but at length, after an hour's persuasion, we succeeded in inducing him to speak, and he answered "yes" when asked if he understood what was said to him. When asked if he would try to speak, he answered "yes," and his face slightly relaxed its cold gaze and became lighted up with a happier expression, which at least signified a willingness to make an effort. When his nurse was pointed to and he was asked who she was, he answered "nurse." He promised that he would make an effort and would speak when again spoken to. It then appeared evident that the organization of the mind was not in any way disturbed, but that its machinery was inactive; there was neither excitation nor depression of any mental faculty; there was no mania, neither was there melancholia, nor any manifestation of disturbance, of emotion, or volition, or of any perversion of sense or intellect. It was a case in which the mind had come to a full stop. The impression of externals failed to set the currents and trains of thought in motion, and the patient might be described as seeing without observing, feeling without appreciating, and listening without hearing, or hearing without comprehending; there was no lesion of attention, but no impressions were made of sufficient strength to produce comparison, and his mind was an almost absolute and complete blank.

This is the condition, to which I apply the expression, Acute Dementia; and from the description of the case I have given, you will, I think, hardly fail to understand the nature of the disorder. You have seen one or two cases at Peckham House, I think you will have no difficulty in recognising it when you see it again.

This affection is by no means so uncommon as at first you might be inclined to imagine; and though the dementia is not always quite as perfect in degree as happened in the boy whose case I have just related, yet it is often quite as sudden in its attack, and it is usually quite as definite in appearance.

I had charge of a boy, by initials G. K., in St. Luke's Hospital, æt. 20, by trade a watchmaker. He was reported to have left his home one morning, and was afterwards found twenty miles away from it, silent, and unable to give any account of himself.

On asking him his name at the time of his admission he did not answer. The next day I again asked him what his name was, and he said he did not know. For several days he either could not or would not speak at all, or else gave only monosyllabic answers. One morning I asked him how he was, he hesitated and said "fairly." I then asked him if he knew where he was, and he said "no." Another patient then came up to me and said that G. K. believed himself to be a horse. I then asked him if he was a horse, and he said, he did not know, but that he thought he was. I asked him how many legs a horse had, he said, "four." In answer to the question how many he had he said, he did not know; I asked him then if he thought he had four or two, he said, it was that, which he could not make out. I then asked him if he had a horse's head, he said, no, he thought he had the legs of a horse and the head of a man. A few days later I found him sitting in a heavy chair, which he was rocking backwards and forwards, and at the same time he was striking his feet on the floor in regular time to imitate the step of a horse in full trot: I then asked him again if he were a horse or a man, he said he did not know but he thought he was a horse. I asked him if he did not think he was a donkey, he said he did not know but he thought perhaps he was. With the exception of this conversation, sustained on two or three occasions, he hardly ever spoke after his admission into the hospital.

The first certificate upon which he was received into the establishment stated that he was gloomy and absent, frequently laughing, but unable to give any reason why he laughed, and that he could not remember why he did so. The second certificate stated that he was in the habit of leaving his home, and that he was frequently found wandering away for miles without being able to give any account of himself. His general health was reported to have been always good. With the exception of an admission of the habit of masturbation, which his brother informed me he had confessed to having practised some time before when he was in America, and the possible exhaustion of the 20 miles walk, his case contained no definite features pointing to an exciting cause. But there was a tolerably definite history of insanity in the lad's family. I learned from his brother that his mother, and also an aunt, had been, as he described them, "queer," and evidently laboured under unsoundness of mind, and the brother himself who applied to me and asked me to take him under my care, stated that he feared that he was going out of his mind himself. He complained that at night-time he was sleepless and excited, and in the day-time he was depressed to such a degree that he was often forced to leave his work, and whilst he was speaking with me, his nervous organization appeared so highly irritable, that I was convinced that a very slight excitant would upset the balance of his mind, and render him a lunatic. He was suffering from mitral disease.

The demented patient, however, gave no evidence of somatic disease; his heart, lungs, liver, kidneys, and spleen, all appeared sound, and he was well nourished. He belonged to a very temperate family, and he had been a teetotaler all his life, his father too had been a teetotaler all his life, and I was unable to obtain any clue to syphilitic or any other deteriorating disease. The attack had continued for eight months when I first saw the patient. His appearance was intelligent, but upon watching him you might have noted that he was unobservant, and without thought. He was very obedient, and would do anything he was told to do, otherwise he conveyed the impression that his mind was a complete blank, which, no doubt, it was.

As a rule, the subject of acute dementia becomes more or less suddenly attacked, he loses his mind, or, in strict acceptation of the term, becomes "out of mind," or his mind is thrown into abeyance suddenly. His mental faculties are not necessarily disturbed, they simply cease to act, and his brain apparently ceases to perform its function of cerebration. It does not appear that the attention is engrossed, or that the mind is concentrated upon some one object or idea, as in melancholia, the only form of mental disease with which you might confound it; neither is it wandering upon one or more trains of thought, as in the condition popularly called absence. The mind of the demented patient has simply stopped working, and if from time to time it becomes at all occupied, it only does so by receiving impressions very passively, and it is incapable of being moved to comparison, or judgment, or reflection, or suggestion; you may compare the condition to that of a machine which has ceased to work, not because it is actually out of order, but because its motive power has been withdrawn.

Though pathological anatomy has furnished but little certain evidence as to the precise state of the brain in the subject of acute dementia, I think that some facts from the clinical history of the disorder may help us to form an estimate as to the nature of the condition, and perhaps furnish us with some reasonable ground upon which we may base our mode of treatment.

In the first place there is neither the excitation of mania, nor the depression of melancholia. The patients become demented, and sometimes, though not always, exhibit phenomena resembling delusions, *i.e.* they give expression to incongruous ideas; as a rule they are merely uncorrected impressions, and fall short of delusions because they seldom become incentives to action. The patients have lost the faculties of correction, comparison, thought, and memory, and so completely do they lose their memory, that they forget their own names. The evidence points to an arrest of brain function, but to an arrest of a more or less passive kind, whatever the cause may have been, and it is highly probable that the arrest depends upon brain innutrition. Its association with heart disease is more common than is generally supposed, and I believe that as we progress in knowledge, and gain facts for associating exact pathological conditions with palpable phenomena, we shall find many of the errors in brain nutrition arising from defects in the circulation, caused by imperfect heart or damaged vessels.

There is no doubt that alteration in the size of the vessels from changes in their walls and occlusions from emboli, often produce more or less dementia; some such cases have been spoken of, and are spoken of by the vague expression—softening the brain. But we also get acute dementia without such coarse conditions, and it is certain that the defective supply of blood to the brain, occurring from any cause, will produce the disorder. That obstructive heart disease was a common cause of dementia and other cerebral disorders, I was convinced from numerous observations; and I found a very complete confirmation of this view, in the admirable lectures delivered in 1843, by Dr. Burrows, before the College of Physicians.

In these lectures, Dr. Burrows mentions Corvisart in particular, who made some observations on the effects of diseases of the heart upon the functions of the brain; and Dr. Burrows gives reports of twelve cases in which mental symptoms were associated with heart disease, one if not two of which, numbers nine and ten, were unequivocal cases of dementia.

The effect of masturbation as an exciting cause is plain from its exhaustive tendency. But it is a practice which, as I have already told you, when indulged in persistently, points to a potentiality of insanity.

You have seen an excellent example of dementia, associated with masturbation at Peckham House. The patient has a peculiar shyness and slyness, withal he rarely speaks; and like many demented patients of his class he eats voraciously and inordinately.

In most cases the effect of exhaustion as a cause of dementia is apparent, but you may see the dependence of the acute condition upon exhaustive influences, even more strongly marked in two cases I will now relate to you, than in those we have considered. In both of them the puerperal state was the direct excitant of the attack; and in one of them the patient's mind had been preyed upon by moral disturbing causes for some time before.

M. S., was a young woman of four and twenty, the wife of a mate of a merchant vessel that had not been heard of for some months, and in consequence, the patient laboured under much anxiety, and fretted greatly. Eight days after her confinement, which took place on the 28th of February, 1870, she gave evidence that she was labouring under delusions; she stated that she had been poisoned, that her child would die also, as it had taken the breast since she had taken the poison, and that if it did not die she would cut its throat, as "life was of no avail." She also said that "they" were throwing her baby out of the window, and in consequence she screamed, and maintained the screaming almost all night. It was necessary to remove her baby from her, and I saw her on the following day. She then made to me, the statements above detailed. I found her sweating considerably, but her lips were parched, her breath was foul, her tongue was furred, and it had an ominous ulcer upon its left side. The lacteal secretion was almost entirely suspended, and she was refusing food.

The next day she was brought to St. Luke's Hospital by her father and a nurse. She said she remembered seeing me the day before, and that the people who were with her had tried to poison her, and that she did not want to go away again with them. She went most willingly into the ward, when she was at once put to bed. On visiting her shortly afterwards I found that her powers of conversation were very limited. She looked at me for some moments, trying to speak. She then burst out crying, and asked me, "Are you a doctor?" She at first refused food, but after a little persuasion allowed herself to be fed with a spoon.

The third day she again endeavoured to give expression to her sensations or thoughts, but she was unable to proceed beyond the delusion of the poison, and the throwing of her child out of window. I encouraged her to speak, but she was quite unable to do so, beyond asking the question, "Are you a doctor?" As I was leaving her room she said, "Don't go, and I'll tell you all." She began to cry, and upon my returning to her bedside she said, "Am I a married woman?" and "Have I had a baby?" but she was unable to say more.

She continued very much in this state for several days, exhibiting an anxiety to speak, but unable to give utterance to more than a set of stock phrases. The excessive sweating abated, the skin became more natural, the breath less foul, and the tongue more healthy, with the exception of the ulcer, which appeared sluggish and inactive. She seemed free from pain, nevertheless she did not understand very much of anything said to her, and was almost incapable of grasping any idea or of answering any questions. She was constantly wet, and passed her motions in bed. After two or three days she said to me one morning, "Am I Lady Mordaunt?" "Are you my husband?" "Am I a married woman?" "Where are my rings?" Her rings had been taken charge of, as she removed them from her finger and dropped them in her bed. A few days afterwards she became depressed, and would not speak at all, and every time I went into her room she either covered over her face or else began to cry. In two or three days more she became very restless, and would not keep in bed unless watched. She then became depressed again, and one night attempted to tear her eyes out. The following morning, and for two or three succeeding mornings, she had an attack of excitement, in which she broke the jug and basin on her wash-stand. The paroxysms, however, passed very quickly away, but left her very much depressed. After three or four mornings they ceased to recur, and she, though very melancholy and frequently giving way to tears, began to occupy herself with a book. She became clean in her habits, and from that time steadily improved in body and mind. She then became free from delusions, and, though at first very weak, began to gain the power of mentally grasping an idea. The melancholy disappeared, and she became cheerful, and completely recovered. The sequel to the story is very sad. The husband had gone to sea soon after his marriage, leaving his wife in the charge of her mother, to whom he gave money for her maintenance. He, through a series of misfortunes, was not heard of for many months, and the funds being exhausted the mother recommended her child to marry again, but failing in persuading her to do so, suggested to her the streets as a source of income, and afterwards encouraged the advances of a lodger, to whose perfidy the poor creature fell a victim. The husband returned while his wife was in the hospital, nineteen months after his departure. The scene which occurred when he heard that his wife had shortly before been confined is not easy to describe, and it was intensely painful. It became my duty to impart the information to him, and I did so in the presence of his father-in-law. A greater moral shock, than for a young man of a generous and noble spirit to return, inspired with hope, after a long and necessary absence, to a young wife, and find her the mother of another man's child and an inmate of a lunatic asylum, could I think hardly occur.

In this case a history of mental and nervous disease was not obtainable from the father, but I was astonished, on the second day after the patient's admission, at seeing the father and with him the mother, who had travelled from Lancashire on hearing of her daughter's alienation.

The father had brought his wife to the hospital, being unable himself to reason with her or to deal with her. She demanded in a most excited manner to see her daughter, and upon my endeavour to point out to her the inadvisability of visiting a patient so soon after admission, she exclaimed, "Oh! I cannot reason." "I am mad myself." "You will have to take care of me." It was evident to me that, if not actually insane, her mind was as nearly as possible off its balance, and that at all events there was a potentiality of insanity in her, and I have little doubt, had I been afforded the opportunity of a more searching enquiry, I should have found clear evidence of hereditary insanity.

The whole history of the case, however, points to exhaustion, and the recovery, which undoubtedly was due to the restored nutrition, confirms this opinion. The patient, descending from a predisposed stock, was first exposed to privation and want, driven to immoral practices in order to obtain food, suffered from remorse during the daily impoverishing influence of pregnancy and starvation, and finally, broke down under the exhaustion of parturition.

A most interesting case was that of a woman named A., aged 32, who was attacked after the birth and loss of her fourth child. The case was in the highest degree remarkable, and presented many aspects which bore comparison with the one already mentioned, which obtained celebrity in the Divorce Court in 1870. At her own home the patient used to wander about the house without object, and would not converse when spoken to, or she would pettishly answer " Don't know." She would not notice her children, and when asked about them would answer " Don't know," as though in these words she expressed all that she was mentally capable of. She did not know the number or the names of her children, and when asked the names of some pieces of money was often unsuccessful in giving them their designation. The calculation of the sum of two or three small coins, as the addition of two sixpences or three fourpences, was a matter of impossibility with her, neither did she appear to recognise a difference between the values of a shilling or a sovereign and miscalled both, and with most extraordinary child-like simplicity she would answer " Yes ?" inquiringly, to anything, but as though she accepted the truth of every statement made to her, however absurd ; I often questioned her, and put ridiculous propositions before her in order to test whether or not she had the power of comparison, or judgment, or correction, but she had altogether lost this power. I asked her one day how many shillings were contained in a pound, she answered " Don't know," I then asked her if twenty shillings made a pound, and she answered inquiringly " Yes ?" I then asked her if she thought eighteen shillings made a pound, and she answered " eighteen ?" " Yes " I then asked her if she thought she was queen of the Moon, and she answered " Don't know," I then asked her definitely whether or not she was the queen of the Moon, and she answered affirmatively " Yes."

This patient almost daily asked me whether I was not her husband, and she used to follow me about the ward and sometimes say, " Surely you are Mr. A." I learnt from her husband that formerly she was very fond of playing cards, but that often in the middle of a game she would become abstracted, and altogether forgetful that she was playing. He stated, however, that after a little while she again awoke, as it were, to the consciousness of her game, and resumed it as though nothing had occurred. At the time she was admitted into the hospital she was too demented to play cards, or in fact to do anything.

She died of acute tuberculosis, which ran its course in fourteen days. She had, during her last illness, some lucid moments, in which she recognised her husband and some of her other relatives, and spoke of and remembered the names of her children. In this case it was maintained throughout that there was not any hereditary taint.

A potentiality was, however, indicated in the existence of epilepsy which was a sufficient predisposing cause, whilst the physical and moral excitement was enough to have developed any form of primary insanity. The condition—acute dementia—is clearly primary, its more or less sudden invasion distinguishes it from chronic dementia, and its prognosis is, as a rule, bad. Some cases recover, but only a small percentage, and encouraging terminations ought only to be looked for in cases where the history points to a very certain source of excitation in an exhaustion which you see some hope of combating with food, rest, and stimulants.

The treatment of acute dementia must necessarily be based upon general principles. The patients in some cases require to be kept in bed, at other times you must seek to awaken their currents of thought by amusements and entertainments, such as music, or dancing, or some other diversion. A good supply of food is always necessary, and in many cases a good amount of stimulant is imperatively demanded.

Chronic Dementia.

We must now consider the condition known as chronic dementia which we shall find differs considerably from acute in many of its characteristics, yet agrees in its main feature of " negation of mind" or loss, in a greater or less degree, of a mind that once existed.

Chronic dementia appears in two varieties, it may be the sequel of acute, as we saw in a case at Peckham House, more often it is the sequel of mania, both acute and chronic; sometimes it follows melancholia, or it may occur as aprimary disease.

As a general illustration of chronic dementia following acute insanity I may relate the case of a woman who was in St. Luke's Hospital for 35 years, who used to attract attention by fresh scratches which her face always bore, and who used to come up to everybody and commence talking to them from the moment they came into the ward. You could not have failed to be struck with the difference between her case and those of the acute insanity which were seen in other wards. The patient used to follow anyone or everyone about the ward, chattering unceasingly the whole time. Her tongue was not only metaphorically, but literally, too long for her mouth, for it protruded at all times about three quarters of an inch beyond her lips. Years ago I understand she filled her place in the society of her day; when I knew her she was a demented old woman. Mania in attack after attack had laid his claws upon her, leaving her intellectual faculties more and more impaired after each seizure, till at length she became reduced to the condition in which I first saw her, viz., a state without intellect at all. She did not even know her own name, and the only sentence I was ever able to understand was certainly a most droll one, for after a very voluble chatter, apparently of a most good-natured tenor, she used to finish with the intelligent expression of "all these crazy people."

Another remarkable case was that of a man, F. H., aged 32, who had suffered from an attack of mania seven years before. He used to sit perpetually in a corner with his head buried between his knees. He could never be induced under any circumstances to speak, and I was only able to trace in his miserable life one act which I could at all recognise as a result of reasoning. He seemed to have an intense horror of chapel, and he was perfectly aware of Sunday when it came round, and at chapel-time on Sundays, he would endeavour to get into a water-closet in order to elude observation, and he used stubbornly to refuse to go down to the airing-court on Sundays lest he should be inveigled into the chapel.

The history of his case is very simple. The man was for some time in India where he suffered from sun-stroke, and the combined effects of sun-stroke and climate, coupled with the exhaustive effects of onanism, which he was said to have practised after the sun-stroke, were followed by an attack of mania; and his exhausted brain tissue had never been able to recover its normal condition; so permanent, apparently, was the brain change in this case that recovery presumably was impossible. His appetite was enormous, and he would devour any quantity of food within his reach unless checked.

The third case is one of singular interest as it occurred in one whose intellect was perhaps more brilliant than many thousands of his own generation.

The patient was an artist of very considerable fame, but his intense interest in his profession became an incentive to excessive labour which reacted as an excitant, threw his overwrought brain off its balance, and he became the subject of acute mania. He suffered and recovered from six or seven attacks, but at length he was again affected, and the last seizure left him in a hopelessly demented condition. And yet even in the midst of the wreck, from time to time, he exhibited some flashes of genius, and the evidences of a master mind became apparent. I often watched him drawing, with a stone, grotesque figures upon the gravel, and yet the poor fellow had not sense enough left to attend to the calls of nature in the manner in which education, at least, has taught the majority of our race. He knew his name, that is, he answered if he were addressed by it, but he could hardly comprehend the most simple question, and rarely spoke or gave a coherent answer.

Perhaps few of the inmates of an asylum appear less interesting than a group of demented patients; and yet the history of their cases is often full of the most touching incidents that could chequer the pathway of ordinary human life.

In the chronic form of the malady the dementia is not always as absolute and profound as it appeared in the cases I have detailed. You will find various degrees of dementia in all asylums. You will observe, too, that chronic dementia is a very common form of mental disease in asylums, it is perhaps more common than any other, and the patients, who are often erroneously called imbeciles, always present a characteristic vacancy of expression, which, to say the least is remarkable. Generally they are listless and apathetic; if they show signs of intelligence it is always childish; often they are obstinate and mulish, and sometimes they are irritable and very easily roused.

When the conditions of dementia have far advanced, individuals of this class usually, are unable to tell their own names. Many are dirty in their habits, and their time is spent in absolute idleness. They will sit, muttering to themselves, or listlessly fidgeting with any object within their reach. Child-like, they will collect together all manner of rubbish; but, purposeless, and as observed by Dr. Daniel Tuke,[*] "without the constructive genius of the child," the toys not of fancy, are gathered to be left; and almost the only propensities ever apparent are those of meaningless violence, or in cases of semi-dementia sometimes you see mischief. These patients often have enormous appetites, and unless strictly watched will devour their meals voraciously and without mastication.

You will sometimes find cases of chronic dementia in private practice, which will cause you great anxiety, and upon which you will have the greatest difficulty in arriving at conclusions.

A short time ago a gentleman called upon me, and detailed part of an endless history of domestic unhappiness resulting from, as he said, his wife's irritable temper and inflexible obstinacy. The consequence being the neglect of his children and the improvident waste of his means, his resources too being by no means great. After years of unhappiness, he said he had at last been led to think his wife must be of unsound mind; and he had been led to the conviction by the circumstances, that shortly before he called on me, she had several times left her home and her children with-

---

[*] Bucknill and Tuke. *Psychological Medicine*, p. 114.

out assignable cause, and without leaving instructions for proper provision for them; and on one occasion, without notice or warning, absented herself for a week, and took up her residence at the house of some relatives. He desired me to call at his house and see the lady, but at the same time he told me that he expected she would refuse to see me. I went, and the moment she heard that there was a stranger in the house, she shut herself up in a room by herself, and replied to her husband, who sent to her and asked her to come down, that she would not see me. The husband expected that she would leave the house and watched for her near the door for about two hours, but she made no movement to go. Afterwards, we went into the garden, and on our return to the house the husband learned that his wife was in a room in the basement; he opened the door and unceremoniously ushered me in, and I found there, sitting at a table, a demented woman, playing with two or three models of horses cut out of paper. She was small, shrivelled, and sallow, with the unhealthy odour I have already mentioned as common among the insane. She immediately became friendly and consented to play the piano; but her music was very incoherent, though I was given to understand that she had been a most accomplished musician. I afterwards gained a long interview with her, but the whole of her conversation was feeble and childish. She admitted that all duties were wearisome to her, even attention to her children was more than she was equal to, and she said she preferred to do nothing; and to leave her children, thirteen in number, to their own devices, or to the tender mercies of servants. The unfortunate patient who was now and again incoherent in her conversation, from time to time, cried piteously whilst I spoke to her; and curiously enough she volunteered the statement that she thought she was demented. Her husband was delighted at the appearance of his wife, and said she was better than he had seen her for years; and it became a painful duty to have to tell him that his wife was suffering from chronic dementia, apparently of years standing.

The early history of the case I was unable to arrive at, but the general facts indicated several attacks of mania. There was an indefinite account of some mental affection in a former generation, but it had been carefully concealed from the husband; the lady's relations were very angry at the suggestion of mental disease, but one of her children has had a definite attack of mania.

The patient has since died. Her dementia became more profound, and she died from pneumonia, and general wasting, apparently resulting from imperfect innervation.

I do not think that pathological anatomy will ever display much more than we know already of the nature and condition of chronic dementia; we often find the affection following upon acute disease, as acute dementia, acute mania, and acute melancholia, and we frequently find after death the ordinary evidence that acute mental disease has existed, we find shrunken and wasted brains, abnormal vessels, abnormal membranes, and abnormal bones. The character of the abnormal tissues indicates an old or confirmed, rather than a progressive change, and like a cicatrix, or a false membrane, often looks like an attempt at the reparation of a damage, as that has been done by an inflammatory or degenerating process, but in the head as in the chest and other parts, nature-mending is necessarily patchwork, and in no case is it possible that the new state can be other than worse than the old. Chronic Dementia is the result of wasting of the surface of the brain, by which its function of thought is suspended. The mental powers may be considered as paralysed, but as the bodily powers do not show the paralysis, it is presumable that the centres of motion in the brain and the cord are unaffected.

I have placed on the table some sections of wasted brain; they are very typical of wasting, though they are from subjects in whom the dementia was not profound; they are not, perhaps, wasted to the same degree as the brains of the insane paralytics which you have seen. The character of the change, too, is different, for the case of the progressive paralytics the brain cells were for the most part filled with granules. Whilst in the sections before you the cells are attenuated rather than granular.

Primary Dementia.

The other variety of *Chronic Dementia* which I spoke of, is perhaps, equally hopeless as the one we have just considered, or perhaps it is even more so. It begins insidiously but progresses with sure and certain strides to its end.

The invasion of the disease is often marked by confusion of thought, and hesitation, and perplexity in expression, rather than any apparent stupidity. A failure of memory, often marked by a forgetfulness of the subject upon which the person has commenced speaking—a forgetfulness of names and of dates unnatural to the person, with often a painful consciousness of the defect which he carefully avoids exhibiting by every means in his power. Sometimes it will be impossible in one or two interviews, to gather from the patient's conversation that there is any departure from his normal state. When, however, the patient attempts to write, the evidence of the condition becomes apparent. You will find that such patients are, firstly, an undue time in composing a sentence, secondly, if a few of the first sentences be plain, the later ones will be confused; and usually the spelling will be very defective.

It seems at first curious, how in these cases, the failure of memory is for the most part with regard to recent events, while occurrences long passed away are as prominent in the mind as though they were of yesterday. A little consideration will, however, furnish us with a reason for this; the brain formerly in health received its impressions firmly and completely, which remained and became the permanent physical condition of the cells bearing them; whilst the brain now atonic and imperfectly nourished has lost its adhesiveness, and is incapable of receiving permanent impressions. The power of reasoning, when within the sphere of recollection, is not in any degree impaired, whilst judgment to be formed out of recent impressions is almost impossible: hence the faculty of attention as regards the immediate cannot be aroused, and by and by inferences from ideas formerly in the mind cease to be drawn. As the case progresses, the patient becomes incapable of comprehending anything that is said, till at last he sinks into the complete or confirmed state of dementia which I have already described. Such patients pass almost an organic existence, scarcely conscious of life, and without either desires, affections, or aversions.

The physical condition of the brain of these cases, is on the outset, probably that of atony, afterwards passing into atrophy.

Very many of these patients become epileptic, as some you have seen at Peckham, and the epilepsy is no doubt associated with the wasting of the brain. Some few of the patients become paralytic. In the early stages these cases are, sometimes, complicated with attacks of transient maniacal excitement and temporary depression, rendering diagnosis difficult; but one of the best distinguishing tests is writing. The subject of incipient dementia, or of incipient primary dementia, as it is sometimes called, soon forgets how to write or how to spell, whilst the maniac will remember. You will hardly mistake incipient primary dementia for melancholia; for although the melancholic is often slow to write, yet you can very readily arrive at the pre-occupation of his mind, which is the real cause of the arrest of the ready movements of his hand and pen.

Sometimes in the early stages the symptoms are recurrent as in a case you saw at Peckham on Friday last.

I lately saw a curious case of incipient primary dementia. A gentleman was said to have had one or two very transient maniacal attacks in which he was noisy, he had also had one or two despondent attacks, when he not only threatened to commit suicide, but in one of them actually made an attempt to cut his throat.

He was a remarkably well-read man, and when he could be induced to converse on some subject of his reading, was regarded by his friends as a charming companion. But when I saw him his memory was seriously impaired, and he would speak of recent events as having occurred a year before. He would lose his way in the neighbourhood he had lived in for years, and he was several times taken charge of by the police as a wandering lunatic. There was some, but not very clear evidence that he suffered from *le petit mal*, and on one occasion, it was said, he had an attack of paralysis which was sudden and transient, and from the history I should judge that it was epileptic. He spent most of his time in bed, and rarely went out except at night time, and habitually turned night into day and day into night. When I visited him at 5 o'clock in the afternoon he was in bed; he complained to me most bitterly of his brother with whom he lived, his grievance being that his brother kept a watch upon him. I ascertained, however, from numerous enquiries, that he could not have been more kindly treated than he was by his brother, who erred rather on the side of indulgence, than of over carefulness, and allowed the demented patient every possible freedom, even to the total upsetting of his household. The brother told me, that he, some three weeks before I saw the patient, had asked him whether he would not like a change to the country, and the patient wrote an answer that he would like a change, and that he would make arrangements for himself; but when I questioned him on this point he had forgotten the circumstance altogether, and complained of being under his brother's thraldom.

Such cases may give you some apprehension; at the termination of your first visit, you may have some doubt as to the patient's unsoundness of mind, because he has spoken to you rationally, but after you have seen him two or three times, you will certainly discover that he is unable to take care of himself, or to manage his affairs.

Senile Dementia.

A form of chronic dementia to be noticed in passing is that which is associated with age, or, as it is called, senile. You saw some cases on Friday last at Peckham House, one in particular, a woman of 96, who was able to speak of circumstances connected with her girlhood, though she did not know the events of yesterday. You could not but have noticed the tremor from which she was suffering and which is one of the accompaniments of old age and wasting of nerve tissue.

The condition bears in its clinical features a slight resemblance to paralysis agitans, but I do not think that pathologically there is any resemblance in the structural changes. Senile trembling does not appear to be associated with sclerosis of cord and pons varolii, whilst paralysis agitans is associated with a peculiar form of sclerosis, and a very characteristic change in the grey matter. Every here and there you will notice in the sections under the microscope, Case no. 13, and which are from a case of paralysis agitans, little groups of granules which have taken the colour very well. It would be premature to speak definitely as to the constitution of these granules, but the beautiful web of areolar tissue which you see is conclusive of formative change; the granulation looks very like retrograde metamorphosis of the cell, but I cannot yet speak with certainty about them.

I recommend you to examine the specimens on

the table carefully, as you will not often get the opportunity of studying the nervous tissue of patients who have suffered from paralysis agitans.

A fact which has often been observed is that the brain, in common with all the other tissues, wastes with age, and dementia as the result of a wasted brain is not therefore surprising; the wonder is that all old people do not become demented, especially when the high pressure living and wear and tear of the present day is taken into consideration. The main features of senile dementia are loss of memory, restlessness, and irritability, and, as you may often observe in asylums, some of the patients become spiteful, and are prone to interfere with those near them, added to which the subjects of senile dementia become suspicious and frequently labour under delusions. A very common delusion amongst the sufferers from senile dementia, arising from a combination of suspicion and loss of memory, is that they are being robbed, and the fancy that they are being robbed of their clothes is as common, or perhaps more common than any other. The subject is one of only passing interest, and is well depicted in the last of Shakespeare's seven ages:

"Sans eyes, sans smell, sans teeth, sans everything."

Infantile Dementia.

A more interesting form of dementia, is that which is not inappropriately called "Infantile" since it has its origin in infants, in whom, it is reasonable to suppose, the brain is healthy at birth, but become deteriorated afterwards. The subjects of this form of disease, may pass through a year or two or more of their existence, show good signs, and give good evidence of brain development, and what is more, of intellectual development, after which the brain power fails. You must distinguish this condition from congenital imbecility. It is in a degree *amentia*, because the mind never had the chance of developing to any degree, but it is *dementia*, because it is the result of disease which has commenced after the intellect has begun to form.

This form of dementia may be more or less sudden, or it may be gradual, and it may be associated with various deteriorations and defects in the nervous organization, as for instance, paralysis, some form of which is very common in infancy. Various forms of paraplegia often appear. Often, we see hemiplegia, amaurosis, and ptosis. Mutism, and deafness too, are common accompaniments of Infantile Dementia. Both the mental, and of course, the accompanying bodily defects are secondary, and the result of changes which have occurred in the nervous substance. Very often the child grows, and the body becomes developed, but the mental powers cease to mature, and in some cases retrograde and become altogether lost. In physical proportions and age, the subject may become adult, yet the mind remains in the condition of infancy.

I had under my care, for a long time, a child, who up to the age of nine months exhibited a fair amount of mental development, when it had an attack of effusion into the ventricles. It is now nearly three years old, is hemiplegic and blind, can only take food by the reflex act of sucking, and it cries when in want of food, otherwise it gives no sign of intelligence, and passes its motions and urine in its bed.

Some cases have advanced so far, as to have learned to speak more than one language, and with discrimination, and then become demented. I have had the opportunity of seeing two such cases through the kindness of Dr. Langdon Down, formerly the Superintendent of Earlswood Asylum; and who, now, has a beautifully-appointed private establishment for the reception of idiots, at Hampton Wick.

The causes of Infantile Dementia are various. Any circumstances inducing arrest or morbid nutrition in the brain may produce it. Drugging with opium or ardent spirits, may so deteriorate the brain of an infant, that atony or atrophy of the hemispheres results, and mental action or cerebration becomes more or less suspended. But any brain disorder occurring in an infant, may result in dementia. What is called, or rather, miscalled infantile paralysis, a pathology commencing with fever, and ending in atrophy, of cord, and muscle, if not of brain and nerves as well, is not an infrequent cause. Effusion of lymph at the base of the brain or into the ventricles, is also a fertile cause of infantile dementia. Infants otherwise healthy, are sometimes born insensate after a very tedious labour, and resuscitated. But the intellect of such infants often developes very imperfectly, and a *post-mortem* examination of such subjects always exhibits some organized lymph, which has been effused at the base of the brain, and around the capillaries. Blows, cerebral hæmorrhages, or any cause which will bring about damage to the infant brain, may result in dementia. Sometimes sun-stroke is the cause, though sunstroke in the infant is very rare, except perhaps in India. Tubercular Meningitis if recovered from, may leave the patient demented. But perhaps the most common cause of all is true hydrocephalus or effusion of fluid into the ventricles, and sub-arachnoid space. Dropsy of the ependyma, as such, does not always remain, the fluid is sometimes absorbed, and its place taken by a connective tissue formation;—a form of sclerosis, first described by Rokitansky, and called by him hypertrophy,—a specimen under the microscope illustrates this condition.

With this state, which has been regarded by some as recovery from hydrocephalus, the dementia is just as common as in true hydrocephalus itself, for the substitution of connective tissue for fluid does not supply healthy brain cells, for the function of cerebration.

Of course you will take note, that fluid effused into the ventricles and sub-arachnoid space and even hypertrophy, does not necessarily give rise to dementia.

There are some notable instances on record in which an enormous effusion has co-existed with tolerable intellectual development. The man Cardinal who died in this hospital, and whose skeleton is preserved in the museum, furnishes a good illustration. But effusion and hypertrophy both produce pressure of the surface of the brain, on the walls of the skull, and although the unossified sections of the skull yield, it is very rare to find serious damage in the head, without some degree of interference with the cerebral function.

The sufferers as a rule do badly, they seldom if ever recover their mental power, and the only cases in which you can express an opinion, other than hopeless, is when the dementia is not great in degree, and when the mental function early exhibits some promise of improvement.

By way of treatment you have but little in your power.

Acute hydrocephalus you must treat actively, and endeavour to induce absorption of the effused fluid. I have found the best treatment, to consist in a bold use of sub-chloride of mercury, and afterwards of bichloride. But unless you manage to make an impression upon the disease very early, you will not do much good.

When dementia in infants or children appears irrecoverable, you cannot give a better recommendation to the friends of these little patients than that they treat them as idiots. They should be placed in idiot asylums, where they seem happier, from the association one with another, than they possibly can be if isolated and at home.

LECTURE X.

IDIOCY AND IMBECILITY.

Idiocy—Imbecility—Heads of Idiots and Imbeciles—Conformation of Mouth and Palate—Development of Brain and Skull—Imbecility—Liability of Idiots and Imbeciles to Mania and Melancholia—Dumb Paralytics often mistaken for Idiots—Diagnosis and Prognosis from Mouth and Palate—Etiology—Syphilis—Fright to Mother during Pregnancy—Drunkenness in patients—Marriage of Consanguinity—Ethnical Classification of Idiots—Treatment of Idiocy and Imbecility.

Idiocy and imbecility, as I have already pointed out, are distinct and definite pathological conditions, but I propose to consider them beside one another as you will always find them associated together, and in fact they require much the same treatment. The distinctive features of the two conditions ought, however, to be clearly defined.

In the condition called idiocy you will recognize at once, that we are not dealing with what may strictly be called disease, but with defective or imperfect development, whilst imbecility contrasts with idiocy in that the brain development may be complete and yet the material of the brain so wanting in normal constituents that mental perfection is impossible in it.

Idiocy does not necessarily imply a less amount of intellectual power than imbecility, though it is usually less. Both conditions are congenital, and, as a rule, the mental defect is rather to be measured by a standard of degree than considered as absolute, for the subjects of absolute idiocy or imbecility do not often long survive their birth.

It is common to see idiots with small-sized heads, and such have received the adjectival appellation "microcephalous" in contradistinction to the class called "macrocephalous," or those cases in which large-sized heads are seen. The macrocephalous cases ought to be classed with imbeciles rather than with idiots.

The true idiot's head measures on an average only about 13 inches in circumference whilst an ordinarily developed head will measure from 20 to 24 inches, or sometimes more. In shape the idiot's head is often imperfect and deformed, the deformity appearing most marked in the frontal and occipital development. You have seen some idiots at Peckham House—two cases in particular must have struck you, the smallness of their heads being so remarkable that you cannot have failed to observe them; the patients are brothers, and are so much alike that they may be taken for twins.

The head of the imbecile is usually fully formed, though you often see considerable contortions and variations in its shape; sometimes the two sides of the face are different, but more often you see distortion of the cranial bones. The most marked features of the distortion are, however, to be found about the mouth and oral cavity.

The lips are usually thick, the thickness being generally more marked in the lower than the upper;

and often they are deficient in muscular power which interferes with their prehensile function, and induces a tendency for the saliva to overflow them. The teeth of the idiot or imbecile are sometimes, but not always, easily evolved. Usually there is a postponement in the evolution of the first teeth, and when evolved, these teeth, often, soon become carious. You will notice that these patients, very often, keep perpetually grinding their teeth, though this grinding of teeth is not confined to idiots and imbeciles. According to Dr. Down's observations, the primary teeth are generally ill-developed and irregular, and their evolution often accompanied with epileptiform seizures. The epileptiform seizures, too, sometimes reappear on the evolution of the second teeth which are usually as irregular and unstable as the primary, and are often, likewise, late in making their appearance.

The tongue is frequently large and long, and its surface is often corrugated into fissures; it is imperfectly under the control of the will, whereby its operation in both speaking and in deglutition become enfeebled.

The most important condition of mouth, however, is that which belongs to the palate. Dr. Down says, that in the large number of measurements he has made, he has, with a few exceptions, observed "a markedly-diminished width between the posterior bicuspids of the two sides, the exceptions being in some cases of macrocephalic idiots who had inordinately large crania depending on hydrocephalus. With the narrowing, the palate becomes vaulted or roof-like in form." "There is very frequently a deficiency in the posterior part of the hard palate from a want of development of the palatal process of the maxillary bone, as well as absence of the palatal process of the palate bone."

So much for physical defect; now let us consider the mental.

We cannot, I think, argue that an idiot is, in the strict acceptation of the expression, a person of unsound mind; his is a condition of amentia, or a state without mind, an absence of mind consequent upon the absence of sufficient, or of sufficiently-developed, brain material, to permit of mental phenomena. Idiots may be almost absolutely wanting in mental power possessing only so much of the central nervous centres as are necessary to allow them to breathe and move. Many of them indeed move automatically: they may be unable to walk but as they sit or lie, their hands move without direction or object, and they are for the most part as unmoved by the appearance of external objects as is a blind man.

Some of these are almost completely insentient. They eat, drink, move, and sleep. They have a corporeal existence but apparently not a mental. The state must be seen to be fully appreciated. The absolute idiot is a being who lives and may be healthy, but he gives no evidence of a mind, and if he has any it does not show itself in this world. But all idiots are not as amental as the description, I have just given, represents them; some possess intellectual power, varying, however, in all manner of degrees, and many of them are capable of considerable education. Sometimes they have one faculty rather than another developed, and this is often more perfect in its degree than any development that the imbecile is capable of.

There are several cases on record, of idiots who have possessed the power of making arithmetical calculations in an extraordinary degree; and lately a remarkable case was mentioned to me. A lad who had not sufficient power of reason to avoid danger, and who was usually listless, though at times mischievous, and always requiring surveillance; yet could calculate correctly almost any number of figures without the aid of memoranda. He was one day asked by a gentleman, who had made a calculation on paper, the number of days that had elapsed since the creation, calculating the date of the creation as 4,000 years before the establishment of the Christian Era. The lad almost instantaneously gave an answer, to which his questioner objected. The idiot then asked to see the memorandum, and pointed out, that his examiner had omitted to take the leap years into his calculation; and this extraordinary intelligence occurred in an individual who would hardly have been able to distinguish between sugar and sand, had they been put into his mouth.

Parchappe, who made very careful inquiry into the subject, concluded that there was a relation between the volume of the brain and the degree of intelligence; and many other observers have since concurred in the general truth of this observation, and shown that the more intelligent the idiot the larger will be his head.

Not only are the brains of idiots small, but they are more or less wanting in the number of their convolutions, and a curious fact, which was first observed by Mr. Solly, is, that although in the perfect human brain the convolutions never correspond on the two sides, yet in the idiot like the monkey, the convolutions of the opposite sides usually do correspond.

It has been made a question, in how far arrest of development in the brain may be the consequence of premature ossification of the sutures of the skull, and the observations of Virchow on this subject are of considerable interest.

He says: "The suture-substance furnishes the material of ossification, the stroma for a deposition of the lime salts, so that under ordinary circumstances, when this bone-originating suture-substance lies on all sides of a skull-bone, it can only increase equally in all directions. But if two adjoining skull bones be soldered together by a premature and complete ossification of the intervening suture, a limit is set to further growth in that direction. If this happens to many sutures at the same time, a microcephalus skull results."

As you well know, the sulci and convolutions of the brain, hardly appear before the seventh to the eighth month of intra-uterine life, and though at birth they are usually mapped out yet their growth and development is to a great extent extra-uterine, and the perfection of their development must be dependent upon the possibility of the expanding brains, and the limit set to the intercranial space. If, therefore, a premature ossification is started, either whilst the fœtus is in utero, or soon after birth, we must expect an arrest of development of the brain; and though the sulci and convolutions may be perfectly mapped out, they will not be of any depth.

Imbecility may be considered amentia, consequent upon an unhealthy condition of brain, commencing whilst the fœtus is in utero; and though the mind may become partially developed, yet it is feeble and abortive. The imbecile often exhibits some degree of cunning, and in him you sometimes find that the animal characteristics, as they are called, are more than ordinarily developed. In particular, you may observe his appetite for food, which frequently resemble that of the brutes. Many imbeciles, apparently, live only to eat, in contrast to the healthy man, who eats in order that he may live. Many imbeciles are, however, capable of learning; and they are, often, particularly apt in acquiring a facility in using tools, and applying them in simple mechanical trades and manufactures. Some of them will learn mat-making, basket-making, tailoring, or shoe-making, with comparative facility, and some of them will accomplish carpentering very well. Some will learn to read and write, others to draw and paint, and some will learn music and even play instrumental music from notes with considerable execution. It is even possible to teach some of them who have never spoken, to speak, when the want of the faculty of speech is dependent upon malformations of the mouth.

All the mental faculties, both sensitive and intellectual, may be somewhat developed; or as stated by Esquirol, sensations, ideas, memory, as well as affections, passions, and even inclinations may exist, though only in a slight and imperfect degree. Nevertheless, the imbecile may be able to think a little, and to judge of the simple acts of life, according to the degree of change and the conditions of development of their brains. Some imbeciles even display considerable shrewdness, and can take care of themselves, but frequently, perhaps generally, they are wanting in moral control. They may even be affectionate to their friends and yet be homicidal, and may murder those they are constantly associated with, and commit the act with a sort of satisfaction and glee. They will sometimes relate their deed without appreciation of its enormity, and show no sign of remorse. Some of them have a dreadful propensity for incendiarism; few comprehend the relations of *meum* and *tuum;* and they are generally very passionate. At one time some semi-imbeciles were in favour as court-fools, because their repartee was thought to be brilliant and witty. Happily, with our advance in knowledge, we are ceasing to admire monstrosities, and to care better for our imbeciles than to take our amusements out of them. Imbeciles as a class, may be regarded as dangerous to society, for a great many of them are so wanting in moral sense, and so unappreciative of moral influence, that they often glide, almost imperceptibly into the class which popularly has received the name of criminal.

Imbeciles and idiots, though they are legally considered as of unsound mind, yet are not mad in relation to development, or in the same sense that a lunatic is mad. A lunatic, or a demented person is of unsound mind, because he has departed from a normal or healthy condition. Whilst the imbecile and the idiot as ordinarily seen, is in his normal condition.

But I should impress upon you that, both idiots and imbeciles, particularly the latter, are liable to the various forms of mental disease, which we have discussed in our former lectures, and you will frequently find that these patients are subject to attacks of mania, and melancholia. Nor is this to be wondered at. Their brains are already imperfect, and predisposed to abnormal action, and they merely want an excitant to throw the little mind they have off its balance. They are not, you should remember, commonly liable to progressive paralysis.

Both imbeciles and idiots are sometimes stunted in their bodily development. The physical frame may, however, assume the adult form and be proportionately developed. Sometimes imbeciles and idiots become hemiplegic, and their physical frame thereby distorted; their muscles then become atrophied or contracted, and their limbs displaced. Sometimes these patients have curved spines, and contorted chests; and of course all these complications tend to render the sufferers more helpless.

There are, however, certain subjects of solicitude, who have no mental defect, but who are often mistaken for idiots or imbeciles.

I have very lately seen two cases. The first was a young lady aged eighteen; she was well up to the age of four, when she had an attack of cerebral hæmorrhage, and ever since has been hemiplegic. She is unable to speak, except to say "come" when she wants anything, and she is unable to stand; some of the muscles of the left side are wasted from disuse and others are contracted. This girl was called an idiot by her friends, and sent away from home. She was half forgotten by her relations, and more than half neglected by those who had charge of her. She was, however, removed to a new home, and I was asked by the lady and gentleman who had undertaken the charge of her, to call at their house and see her. To my surprise I found not an imbecile or an idiot, but a remarkable intelligent girl. The paralysed side of her face was expressionless, but the healthy side was as bright and intellectual looking as one could wish to see. I had a long conversation with her, she understood perfectly all I said to her, and by signs conveyed to me her wishes. She had never been taught to read or to write, for nobody had ever taken the trouble to instruct her; but she was of opinion that she would have no difficulty in learning, and she was desirous of making a commencement. She had taught herself to draw, and was able to draw very nicely. Her special wish, however, was that an effort might be made to restore her left hand sufficiently to use to enable her to hold a knitting needle.

In anticipation of my visit, she had communicated by signs a number of questions she wished to ask me, and when these were put to me in her presence, she nodded assent to each, with a smile and an expression of satisfaction.

I learned that her power of observation, was very great indeed ; not a circumstance took place within her sight or hearing, that she did not relate by signs, to the lady who had charge of her. On examining her mouth, I found no malformation of teeth or hard palate, but the soft palate was stretched tight, and narrowed the opening into the pharynx; deglutition was therefore difficult, but she took time to feed herself, and never put into her mouth a larger piece of food than she knew she could swallow. Her appetite, too, was very good, so that her meals occupied her rather a long time, but she suffered no further discomfort from the palatal condition.

It is probable that the condition of the soft palate might become a bar to her speaking much, but I have no doubt that, with judicious training, she might be taught a few distinct utterances.

Here was a sad instance of a hasty and incorrect judgment, by which an intelligent child had been consigned to neglect, because it was thought to be impossible to benefit her. In her new home she was quite happy, and if she now be allowed the education she desires and needs, it will go a long way to mitigate the seeming cruelty of her paralytic affliction.

The other case is that of a little girl, æt. 4, who is at the present time under my care.

The history of the case points to an effusion of lymph at the base of the brain, but the child has no lesion of intellect. The patient is unable to walk from contraction of the gastrocnemii and other muscles, but I expect that she will attain to the power of walking after she has undergone an operation which has been proposed. She is unable to speak, and has her soft palate stretched tensely across the pharynx and constricting its opening, but she seems to have no difficulty in swallowing. This child understands perfectly well what is said to her, always wears a happy expression of face, and has never given the least sign of a want of intelligence, yet because she is unable to speak and unable to walk, her relatives were advised to place her in an Idiot Asylum. Fortunately for the child, she has a most affectionate mother who will not part with her.

In an interesting paper, read in 1871, before the Odontological Society, by Dr. Down, "on the relation of the teeth and mouth to mental development," the author refers to the fact that the frequent imperfection of speech in idiots and imbeciles is popularly regarded as arising from malformation of the palate, and he pointed out that in the imbecile the defect of palate was due to congenital conditions, and that in such children ideas are so little formulated that language is not needed. The defect of palate and mouth, which Dr. Down has urged as strong confirmatory proof of imbecility, consists, as I have already told you, of a narrowing in width and an imperfect development of the hard palate, and great irregularity of the teeth, and it differs very widely from the two cases I have detailed, in which the defect is in the soft palate, and appears to arise out of the paralysis and contraction of the muscles with which the subjects are affected. In the paper I have mentioned, Dr. Down has illustrated the value of the evidence afforded by the physical conformation of the roof of the mouth, in determining whether the defect is congenital, or whether it has occurred in extra-uterine life.

Dr. Down observed that "in children whose idiocy is accidental, arising from causes operating after uterine life," (the class of cases which I described under the head of Infantile dementia) there is but slight deviation from normal condition in the state of the mouth and teeth, while it is in those whose malady is congenital, especially when arising from causes operating at a very early period of embryonic life, that the deviation of the mouth and teeth from a normal condition is most pronounced."

In the same paper, Dr. Down also pointed out the value of observation in regard to defect of the oral cavity as a guide to prognosis. He says, "In children who exhibit any want of mental power, or present anomalous moral or intellectual symptoms, no more anxious question is suggested than that relating to the future of the case. The disposition of property, and other family arrangements, depends a good deal on the answer which is given. We have learned by experience this important fact, that the child who has been born with defective intellect is more susceptible of improvement by physical and intellectual training than the child who has been born with full possession of his brain power and has afterwards been deprived thereof. In other words, that of two children who are the subjects of solicitude, other things being equal, there is a greater probability of improvement for a patient with an ill-developed, than one with a damaged brain. Often it happens that a microcephalic idiot, about whom the inexperienced would entertain no sort of hope, will far outshine, under intellectual training, the fine well-developed boy, the membranes of whose brain have been the subject of inflammatory lesion, and about whose capillaries lymph has been inextricably effused. An appeal to the condition of the mouth is all important in determining whether the lesion, on which the mental weakness depends, is of intra-uterine or post-uterine origin. In the event of the mouth being abnormal it indicates a congenital origin, while if the mouth be well-formed, and the teeth in a healthy condition, it would lead to the opinion that the calamity had occurred subsequently to embryonic life."

I must add to the observation of Dr. Down the note, that you may have, in the non-congenital patients and subjects of infantile dementia, a contraction of the tensor and levator palate in common with other muscles, and you must be careful not to let this point mislead you in your diagnosis.

Etiology.

When we attempt to solve the problem of the cause of idiocy and imbecility, we are at once met with difficulties which are all but insuperable.

The causes which get most credit for producing idiocy are marriages of consanguinity, hereditary predisposition, drunkenness in parents, frights to the mother during her pregnancy, and syphilis.

If we start with syphilis as a cause, the evidence is rather against the probability of its influence than in favour of it. From the able researches of Mr. Jonathan Hutchinson, of the London Hospital, we have an excellent help to diagnosis in questions of the transmissions of syphilitic influence, from certain conditions of the teeth, and a chronic inflammation of the cornea. You are perhaps familiar with syphilitic teeth, but I hand round drawings,* copied from some of Mr. Hutchinson's examples which cannot fail to strike you. The

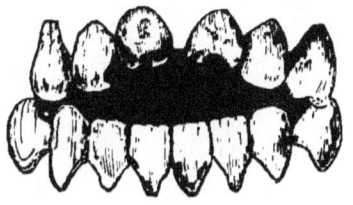

* I am indebted to Mr. Hutchinson, for the woodcut inserted above.

principal are the notched incisor teeth which you see in the drawing, and which, when you notice in your patient, you may accept as positive evidence of inherited syphilis.

In investigating the question of syphilis as a cause, Dr. Down writes that he found very few instances of syphilitic teeth amongst the numbers he examined, and when he did find them he always found confirmatory evidence of inherited syphilis in an associated keratitis or chronic inflammation of the cornea; and Dr. Down has thereby been led to the conclusion that syphilis is by no means an important factor in the production of congenital mental disease.

The evidence of fright to the mother during her pregnancy seems to be confirmed by so few authentic instances, that like syphilis it must be omitted from among the important causes.

Drunkenness in parents, may have more influence, and it has been strongly insisted upon by Dr. Howe of Massachusetts, and his evidence is by no means meagre, but it leads to the subject of heriditary origin of degeneracy, of which the evidence is very pointed and strong. "Because the fathers have eaten sour grapes, therefore are the children's teeth set on edge" is a truism which may be confirmed, at least in some degree, in the study of the cause of idiocy and imbecility. Drunkenness itself is often a symptom only of nervous degeneracy, which in its turn, is the legacy of errors in a generation before. Dr. Down has noted, that in the valleys of Piedmont, it is easy to trace in the cases of the cretans, a gradual degeneracy from grand-parents to grand-children, and he suggests that the same gradual deterioration might probably be traced in idiots, were we able to make such observations as the measurement of the palate and mouth, in two or more generations. The facts of such observations would be most valuable, but I think we have very perfect evidence of the gradual degeneracy, in many well-authenticated cases; one of which, as quoted by Dr. Maudsley in his admirable address to the psychological section of the British Medical Association at Birmingham in 1872, I give you here.

While the Reign of Terror was going on, during the first French revolution, an innkeeper profited by the critical situation, in which many nobles of his commune found themselves, to decoy them into his house, where he was believed to have robbed and murdered them. His daughter having quarrelled with him, denounced him to the authorities, who put him on his trial, but he escaped conviction from lack of proof. She committed suicide subsequently. One of her brothers had nearly murdered her on one occasion, with a knife, and another brother hanged himself. Her sister was epileptic, imbecile, and paroxysmally violent; her daughter in whom the degenerate line approached extinction, became completely deranged, and was sent to an asylum. The degeneracy may thus be traced through three generations, but probably had its starting point in one or two generations before. We see, however, in the

1st generation. An absence of moral sense displayed in murder and robbery.

2nd generation. Suicide, homicide, epilepsy and imbecility.

3rd generation. Mania, and probable extinction of the race.

But to go a little further back, to the question of the starting of the degeneracy, I think that the strongest evidence we have is that which points to breeding in and in, as a cause.

I wish you clearly to understand me on this point. I do not speak of breeding in and in, as *the* cause of degeneracy leading to idiocy, but I speak of it as *a powerful* cause, and in any given number of cases, one which would become a large factor.

The facts were commented upon, in an able paper in the *Edinburgh Medical Journal* in 1865, by Dr. Arthur Mitchell.

The great objection which has been raised to the various statements tending to confirm the belief that frequent inter-marriages lead to degeneration, is that in the breeding of blood-animals, particularly race horses, breeding in and in, is resorted to as the means by which improvement in the race is brought about. The question has been warmly discussed over and over again, and you will find a most interesting essay on the subject, by Dr. Gilbert Child, in the *Westminster Review* of July 1863, or reprinted in his book,* and you will also find in the *London Hospital Reports* for 1866, a paper by Dr. Down, in which he reviews

* *Essay on Physiological Subjects*, 1869.

the opinion of Dr. Gilbert Child, and Dr. Duvay of Lyons; and to this paper I may briefly refer. Dr. Down, who had exceptional opportunities of studying this question at Earlswood, admits that the influence of intermarriage is very great, and he disagrees with the assertion of Dr. Gilbert Child, that "marriages of blood relations have no tendency *per se*, to produce degeneration of race." Yet he seems hardly to agree with the assertion made by Duvay of Lyons, "that in pure consanguinity, isolated from all circumstances of hereditary disease, resides, *ipso facto*, a principle of organic vitiation."

Dr. Howe, who was quoted by Dr. Child, brought forward some remarkable statistics, showing that about half of the issue of marriages of cousins were idiotic, whilst Dr. Child has stated, that he believes the statistics which find half the children of the marriages of cousins idiotic, are drawn from unfair cases; Dr. Down agrees with Dr. Child on this point, and brings forward statistics of his own, and compares them with those of Dr. Howe. I give them as under:

Dr. Howe's statistics. Marriages of cousins, 17 *marriages produced* 95 *children, i.e.* 5·58 *each.*

Of the 95 children 37 were of tolerable health
    1 was a dwarf
    1 was deaf
   12 were scrofulous or puny
   44 were idiots
  Total 95

Thus more than 46 per cent. were idiots.

Dr. Down's statistics. Marriages of cousins, 20 *marriages produced* 138 *children, i.e.* 6·9 *each.*

Of the 138 children 75 had average health and intellect
11 were consumptive
8 were still-born
4 died from convulsions or fits
2 were hydrocephalic
7 died young from infantile complaints
6 were puny and delicate
25 were idiots
─────
Total 138

The proportion of idiots here, as stated, was only a little more than 18 per cent.

But I think that the mode adopted by Dr. Down in reading these statistics, hardly shows the grave deterioratory influence of marriages of consanguinity; for if we consider the question of the number of beings with ordinary health resulting from these marriages, we shall arrive at a very different and striking result: let us take the same statistics.

Firstly, from Dr. Howe's figures we have 95 children, of whom 37 were in tolerable health or a percentage of 38·9.

Secondly, from Dr. Down's figures we have 138 children, of whom 75 were of average health, or a percentage of 54·3, so that there is not a very great difference between the two sets of cases, cited to prove opposite facts, whilst all the instances of other than good health may be looked upon as degeneracy.

In comparison, I have collected 17 families of non-consanguineous marriages, taken from the ordinary community, without selection. The 17 families produced 79 children, or 4·64 : to each.

Of these 79 children,

    68 had ordinary health
     2 died of infantile diseases
     1 suffered infantile paralysis
     2 were consumptive
     2 were scrofulous
     2 were insane
     1 was a deaf mute
     1 was idiotic

Total 79

The percentage of idiocy being thus little over 1·2 per cent., whilst the percentage of ordinary health is as high as 86.

In the face of such facts I think we can no longer doubt the powerful influence of consanguineous marriages in producing degeneracy.

But idiocy is not by any means the only result of breeding in and in : moreover, all forms of mental degeneracy crop up among the progeny of consanguineous marriages, and at various ages. Dr. Maudsley has collected some valuable information on this head. In his book, *The Physiology and Pathology of Mind*, he quotes Morel, whose careful inquiries have brought out some striking facts as to the retrogression of the mental power, and who says of marriages of consanguinity, that they produce morbid varieties of the race, of which he considers madmen to be one ; and he says, further, that the evil unless counteracted, increases through generations, until the degeneration has gone so far that continuance of the species is impossible. He relates the case of the elder branch of an ancient county family, untitled, but prouder of its simple squirehood than the younger branch with its title; it has for generations married only with the members of another great family in the same county. What is the result now? The present representative has three sons, one of whom is deaf and dumb, another is epileptic and nearly imbecile, and the third is scrofulous and feeble in mind and body. The next generation will doubtless witness the extinction of this proud and ancient line.

Morel sums up the history of such cases from the generation in which the first departure occurred.

> In the first there is alcoholic excess, immorality, mental degradation.
> In the second, hereditary drunkenness, attacks of maniacal excitement, general paralysis.
> In the third, sobriety, hypochondria, melancholia, systematic mania, homicidal tendencies.
> In the fourth, feeble intelligence, stupidity, first attack of mania at sixteen, transition to complete idiocy and probable extinction of the family.

Whatever other causes bring about degeneracy of the race, it is clear that breeding in and in is a powerful cause; and if its effect is not seen in the first generation, it can hardly fail to manifest itself in those which succeed.

But I have not yet stated quite the whole case, and as much as I feel that marriages of consanguinity ought to be discouraged, I must give the supporters of these marriages their due. Pure consanguinity appears to produce intensification and refinement, and to perpetuate qualities either good or bad in an exaggerated form; hence it is possible, nay probable, that if the breeding stock of a family could be selected, as is the case with race horses and pedigree cattle, and none but untainted and perfectly healthy members of that family were allowed to breed together, their progeny would be as perfect as the progeny of our blood-animals. As the case stands, however, blood-relations both tainted and untainted, are allowed to breed indiscriminately, and the natural consequence of breeding in and in, when defect exists, is degeneracy.

Before leaving the subject of etiology I ought to mention tuberculosis as one of the recognised causes.

Ethnical Classification of Idiots.

Dr. Down has proposed an ethnical classification of idiots, and it is highly interesting, not only from its practical utility, but also from its bearing upon the question, whether the various races of men, have sprung from a common stock, or whether they have originated in various centres. The evidence furnished by degeneracy, is always a strong argument, whether in plants or animals; degeneracy always retrogrades towards the original type, and this fact gives force to the conclusions which may be drawn, as to a common parentage, from Dr. Down's classification. He has, of course, mentioned that there are numerous representations of the great Caucasian family. But he has found marked examples of the Ethiopian variety presenting the characteristic malar bones, the prominent eyes, the puffy lips, and retreating chin. The woolly hair has also been present, although not always black, nor has the skin acquired pigmentary deposit. The examples have been specimens of white negroes, although of European descent.

Some idiots arrange themselves round the Malay variety, and present in their soft, black, curly hair, their prominent upper jaws and capacious mouths, types of the family which people the South Sea Islands.

He has also found idiots, with shortened forehead, prominent cheeks, deep set eyes, and slightly apish nose,—the analogues of the beings who originally inhabited the American Continent.

But it is the representation of the type of the great Mongolian family, which has particularly been commented upon by Dr. Down, who states that "a large number of congenital idiots are typical Mongols. So marked is this, that when placed side by side, it is difficult to believe that the specimens compared, are not children of the same parents."

The following description of a Mongolian idiot, I quote at length from Dr. Down's paper.

"The hair is not black as in the real Mongol, but of a brownish colour, straight and scanty. The face is flat and broad, and destitute of prominence. The cheeks are roundish, and extended laterally.

The eyes are obliquely placed, and the internal canthi more than normally distant from one another. The palpebral fissure is very narrow. The forehead is wrinkled transversely from the constant assistance which the levatores palpebrarum derive from the occipito-frontalis muscle in opening the eyes. The lips are large and thick with transverse fissures. The tongue is long, thick, and is much roughened. The nose is small. The skin has a slight dirty yellowish tinge, and is deficient in elasticity, giving the appearance of being too large for the body."

Dr. Down adds: "The boy's aspect is such, that it is difficult to realize that he is the child of Europeans, but so frequently are these characters presented, that there can be no doubt, that these ethnic features are the result of degeneration. Dr. Down further remarks that: The Mongolian type of idiocy occurs in more than ten per cent of the cases which are presented. They are always congenital idiots, and never result from accidents after uterine life. They are, for the most part, instances of degeneracy arising from tuberculosis in the parents. They are cases which much repay judicious treatment. They require highly azotised food, with a considerable amount of oleaginous material. They have considerable power of imitation, even bordering on being mimics. They are, humourous, and a lively sense of the ridiculous, often colours their mimicry. This faculty of imitation, may be cultivated to a very great extent, and a practical direction given to the results obtained. They are usually able to speak; the speech, however, is thick and indistinct, but it may be improved very greatly, by a well-directed scheme of tongue gymnastics. The co-ordinating faculty is abnormal, but not so defective that it cannot be greatly strengthened. By systematic training considerable manipulative power may be obtained.

The circulation is feeble, and whatever advance is made intellectually in the summer, some amount of retrogression may be expected in the winter. Their mental and physical capabilities are, in fact, directly as the temperature.

The improvement which training effects in them is greatly in excess of what would be predicted if any one did not know the characteristics of the type. The life expectancy, however, is far below the average, and the tendency is to tuberculosis, which I believe to be the hereditary origin of the degeneracy."

Treatment.

The treatment of idiocy and imbecility has, during late years, been studied with the philanthropic object of ameliorating the condition of the sufferers, and much success has attended the effort.

There is now little excuse for leaving the imbecile or the idiot to live the life of a brute, or to be an absolute burden, taxing one at least, or perhaps more healthy or sound lives, in order to maintain his miserable existence. In a very large majority of cases these patients can be benefited, and they ought at least to have the opportunity afforded them.

The first step towards improving their state lies in association. These patients must be associated together, at home, or amongst those only who are of ordinary mind; they do not feel an influence of sympathy, but when several imbeciles or idiots are together they seen to sympathise with and to companionate one another. They next require good feeding and nourishment; they must then be taught, and the teaching requires judgment, study, and experience, and must be modified to suit each particular case. These patients have to be weaned from bad habits, and trained to observe the decencies and necessities of social comfort; they may then be in a condition, to appreciate and profit by more intellectual culture.

Dr. Maxwell,[*] at one time the Superintendent of Earlswood Asylum for Idiots, writing of the idiotic cases, which had been admitted under his care, observed: "I think they have all improved more or less. Kind treatment, good diet, and attention, will improve the most hopeless cases."

"Many that come in dirty and irritable, not only become cleanly, but get to speak intelligently, to dress themselves properly, and to make themselves useful. Other cases will do a great deal in the school; for instance: we have a case which came in spiteful, obstinate, and unable to read and write. Now he reads well, writes well, also writes from dictation, draws very nicely, can sing several

[*] Quoted by Bucknill and Tuke.

songs, plays on the harmonium, and can drill, which has made him walk upright. He has latterly been in the mat-making shop, and can make the best part of a mat. Another boy has improved in all the above, and is learning mat-making. He possesses, perhaps, the most intellect of any of the boys : but I cannot say that I think he will ever be like an ordinary person. The cases *most favourable* are those between seven and twelve, which are healthy, can speak, and are free from fits and paralysis."

Dr. Langdon Down who succeeded Dr. Maxwell at Earlswood, has comfirmed this statement, and has shown that infinite gain may result to these patients from association one with another. Dr. Down's private establishment is a model of excellence, good taste, and good order. I am indebted to him for his kindness in allowing me to visit it, and observe his cases clinically. I found all the inmates happy, and to a surprising degree orderly; some of them were engaged in fancy work, some in gardening, others can read, write, and draw, and one has been able to made considerable progress in French. The most interesting case of all had learned to play the piano, and favoured me with some music, which was performed with remarkable acuteness and execution.

Dr. Down's experience is, that the cases when associated together never get worse; and the general rule is that the companionship of each other is so congenial, that if any case does not improve under its influence, such case is exceptional, and that exceptional cases are very rare.

My own observation leads me to the conviction that many more idiot asylums are needed, or that every county asylum, should have a building with grounds set apart, for the treatment of imbeciles and idiots.

The ordinary lunatic asylum, is unfitted for the reception and training of idiots, and the association with lunatics does not tend to good, whilst with the knowledge we possess of our power of teaching this sad and afflicted class, and rendering their lives in some slight degree happy and enjoyable, we are acting inhumanly if we leave them in the slough of the workhouse, where, too long, they have been neglected by reason of the parsimony of over-zealous parish authorities.

LECTURE XI.

Etiology of Insanity.

Disease of Mind dependent upon Morbid Tissue—Hypothesis of Possession of the Devil—Moral Influence—Waking Hallucination—Religious Impressions—Potentiality of Insanity—Drunkenness—Precocity—Influence of marriage among the tainted—Syphilitic Diseases of the Brain—Starvation—Blood-supply to the Brain—Narrowing of Vessels—Aphasia—The Development of Language—Chloroform—Railway Travelling—Excess of Brain-work—Influence of Emotions—Disappointment—Surprise—Conscience—Artificial Condition of Living—Influence and state of Society—General note on the Pathological Anatomy of Insanity—General note on the Treatment of Insanity.

We have examined the various forms assumed by mental disease, and we have considered the causes in detail, but it is still necessary for us to consider the etiology of insanity as a whole, or in broad generalities. I should remind you that all the phenomena of insanity are secondary, and that we must remember to avoid the error of regarding disease as having an existence *per se*; we have perverted function in consequence of morbid tissue, but the diseases of mind, which we designate as mania or melancholia, have no existence apart from the tissue the morbid function of which they are the manifestation. The hypothesis of the identity of "the possession of the Devil," and insanity of mind was long supported by the erroneous view which considered it apart from matter; the hypothesis has hardly yet entirely exploded. Not long since a medical man told me that he believed fully that at the present time "his Satanic majesty" actually entered into and influenced the minds of men. This reminds me of a patient who was in St. Luke's Hospital—a girl who believed herself possessed of devils. Her strength was enormous at times, though she is thin and fragile, and wasted almost to a skeleton. In endeavouring to quiet her in a paroxysm of excitement I injected some solution of morphia into her right arm; she told me the next day that I had taken the devil out of that arm, so I suggested his extraction from the other, which a similar injection accomplished. I mention this to show the absurdity of such a theory in practically dealing with the physical conditions we have been considering. In the case of the girl I have just mentioned, the unequal dilatation of the pupils bore testimony to the fact that, in her case at least, the physical conditions of disease were associated with her spiritual belief.

There can be no doubt about the enormous power of what is termed the moral influence, and of the influence of religion over the mind of man, but, at the same time, neither religious beliefs nor moral causes are in themselves sufficient incentives to maniacal acts. The man who dreams a dream, and believes he has had a revelation from God, may not be mad. Ignorance is quite sufficient warrant for many persons to believe in the tangibility of an absurd impression which has floated before the fancy during semi-sleep, but if any impression becomes of such prominence that it excludes all others, and is thereby incapable of correction, then that person is insane, and not unlikely to perform unreasonable and irrational acts. If a person has a dream in which the impression, that it is the Divine will that he shall throw his children out of window, becomes imprinted upon his brain, he is not necessarily mad; his reasoning power on waking may be sufficient to correct the impression, and as long as this is so he is sane. If, on the other hand, he cannot correct the impression but proceeds to commit the homicide in consequence of a belief in a Divine authority for the act, he cannot be considered other than mad. This, of course, involves the question of morals, which we shall consider briefly in its place. The principle of morals involved in the question of sanity or insanity is, nevertheless, the same whether we have an innate sense of right and wrong, or whether our standard of right and wrong be dependent upon impressions inculcated in education, and upon a knowledge of the laws of our country. It really matters little in the consideration of responsibility whether the being has an abstract knowledge of right and wrong; the question is, or should be rather, whether he has the power of choosing right even if he knows it, or of resisting an impulse to action. Such an unreasoning mind as that which cannot correct a waking hallucination, or a religious or moral mis-impression must be the outcome of an unhealthy and imperfectly acting brain. Waking hallucination in a healthy subject results from an imperfection of cerebral function during the waking act, which is immediately corrected when the person becomes fully awake; and the utmost degree of change that powerful moral and religious influence can work in a healthy brain is exhaustion, and though, in a state of exhaustion, a person may be induced to do a foolish act, which he would regret after he had had time to rest and think about it, exhaustion could hardly, under any circumstances, be so great in degree as to allow him, without an attempt at correction or resistance, to be persuaded, or to persuade himself, into the commission of a gross act such as murder. When waking hallucinations, or a religious impression, incites unreasoning acts, they do so because the brain is unhealthy, and the impressions forming the person's sense or knowledge of morals cannot be aroused sufficiently to correct the incongruity.

There is a belief, and it has lately gained credence, that the education and religion of the present age predispose to insanity. But it does not represent the exact fact. That the social condition of the present age, in the more civilized countries, is such as to favour the development of insanity is undoubtedly true, but it only developes mental derangement in the subjects of a potentiality. The higher-class education, of men in particular, and the storm which has long been raging among the various religious opinions of the present age, have together driven thousands on to the sea of doubt, and forced apprehensions, fears, and anxieties, upon numbers who, in a former day, would have plodded on in the even tenor of their ways, with the religion and belief that their mothers had taught them. Thousands battle through the storm which education and religion have wafted over the social horizon, all have more or less felt its influence, and some have been profoundly affected by it, and it has become the excitant which has overturned the balance in the reasoning power of many, who, otherwise, might have passed an unperturbed existence. But it has stricken only those who were predisposed to morbid affection, and the bulk of men pass through the storm unscathed. They mostly assume one or other of the many current beliefs, whichever presents itself as most assimilable with their general impressions, and they regulate their conduct in life rather by their moral sense than by their religion. Among such, you will find numbers in whom, except from a direct physical cause, insanity is impossible.

In our introductory lecture, I told you that a question, which had occupied the minds of numerous observers of mental phenomena, was whether or not a potentiality of insanity was necessary before a person could become insane; and I explained the term "potentiality" as a possibility peculiar to the individual, or a quality which exists *in potentia*. That religious, or scientific, or any other teaching, should *ipso facto* produce, in a healthy brain, the physical change upon which insanity depends is impossible, and we must, therefore, look beyond, and seek for a peculiarity in the individual which separates him from the ordinary community, and renders his brain liable to morbid cerebral action under the pressure of a strain in thought.

There are thousands, nay, hundreds of thousands of families, no member of which could ever become insane, except from a disorganization of the brain, set up, either by such injury or accident, as a blow on the head, or a sun-stroke, or resulting from diseases affecting the circulation, such as morbus cordis, or changes in the vessels; in other words, there are people who have no predisposition to insanity, and who are, therefore, never likely to become insane from primary disease of the brain. But there are other families in which, any member may become insane, and in which, some members do become insane upon a very slight exciting cause; or persons who are liable to become insane from primary disease of the brain, the tendency to which was laid down in the germ. I know a family in which an amental epileptic appeared in one generation, and indefinite, yet fatal nervous disease was recorded in the next. Of the representatives of the third generation, one member died insane at 40, another died of phthisis at 38, and a third is now under my care the subject of epilepsy, which appeared at the age of 35. This patient has four children, one of whom became epileptic at 16. The others appear healthy, but the liability to nervous disease in any of them is very great. From the antecedents of their family the possibility of epilepsy or insanity in any of them is the maximum. In other words, we have in their cases evidence of a potentiality, but whether it ever becomes an actuality is altogether another question.

The great question in regard to potentiality is, of course, its origin, and there can be no doubt about the fact, that in many instances, this origin dates from a remote time, as seen in hereditary cases. It is absolutely certain that much of the insanity of the present time is due to the errors of a former generation, and it is an appalling fact, but a sad truth, that the sins of the fathers are visited upon the children for the third and fourth generation.

Blows, shocks, and sun-stroke, are undoubtedly primary causes of insanity. It has been calculated, that, sometimes, as many as 10 per cent. of the cases in asylums may be traced to such injuries as blows on the head. This percentage is high, but there is certain evidence that blows on the head give rise to progressive changes of the bones and membranes, and that their disorganization often extends to the brain. The acute symptoms which follow a sun-stroke are generally more rapid, in their development than those which accompany insanity following a blow on the head; the course, too, of the former is usually more rapid, and the termination more often fatal. Many of the cases, however, become completely demented, and should the affection assume the chronic form the patients may live for an indefinite time.

Drunkenness is, by many, considered as a primary cause of insanity, and it is certain that there are many cases in which the only history obtainable commences with inebriety, and as we are unable to trace back any cause in a generation before, it is argued by the supporters of the opinion, that we are bound with our present knowledge to place intemperance among the primary causes.

Although it is true that drunkenness is very frequently the exciting cause of an attack of insanity, yet the incentive to drunkenness is the result rather than the cause of brain deterioration in the first instance; a potentiality of insanity is often discovered on examining the history of the persons who complain of constant sense of depression, and who seek to relieve that depression with stimulants. That they obtain a temporary relief from the stimulants is certain; but the remedy is worse than the disease, the stimulant effect of the alcohol having passed away, the depression is increased, and the patient is driven again to seek relief in stimulants, but he finds larger potations are necessary, and intemperance results. It thus appears, that a drunkard may begin to drink because he has the nervous conditions necessary for insanity, and in seeking to relieve his depression sets a light to a train of nerve-disorganization. Opinions I know are much divided on the subject, but I incline to the view that excesses are to be regarded in the light of symptoms rather than causes. One thing is certain, viz., that what is moderation in one man is excess in another. Some men are capable of imbibing an astonishing quantity of alcohol with, apparent, perfect impunity.

There are numbers of men, and some of them, men of great intellect, who drink, what might be described as excessive quantities, and yet, never get drunk, and never suffer brain deterioration in consequence. You will probably come across many such cases, as I have, and they will tend to confirm you in the view, that insanity cannot occur without a potentiality, even when alcohol becomes a cause.

Nevertheless, we have before us the stubborn fact, that in the coal districts, the admission of men into some of the asylums, (in particular the Glamorganshire, as recorded by Dr. Yellowlees), declined to one half during the recent strikes, when the colliers could not obtain drink, and it immediately rose when they returned to work, and were again able to procure alcohol.

About a year ago there was a patient in Stephen Ward, under Dr. Wilks, who told me that he had spent five years of his life in India, and that during that five years he had been dead drunk every night; nevertheless, he had mastered a number of Eastern languages in the time, and obtained first class certificates in the examinations he had undergone, for the Civil Service, in India, where his superior attainments had always gained him employment. He was suffering from partial paralysis when he was in the hospital, but his intellect was in no way affected.

Much that I have said in regard to alcohol, applies with equal force to onanism, which, when persisted in, is in my opinion the result, rather than cause of brain disease.

We have already seen that it is very commonly practised by children, but the question may fairly be asked, does the indulgence grow to a deteriorating habit unless some underlying cause be lurking in the nervous organization? It is a subject well worthy of all the observation that can be given to it; but the reluctance of friends to admit the existence of insanity in their family history, often proves a barrier in the way of investigating this question.

The child I told you of in a former lecture, whom I noticed placing herself in salacious positions, furnishes us with a striking illustration of nervous irritability and morbid sensibility, and I have no doubt that the cause of such irritability, always arises in original defect or weakness of cerebral tissue, or in sympathetic imperfection.

The condition bears comparison with hyperæsthesia, and it may be regarded as a sort of hyperæsthesia of the sexual nerves. Dr. Maudsley, quoting Morel, relates a similar case, in which the lady afterwards became a nymphomaniac.

The excitement of the sexual organs in persons inclined to nervous deterioration, seems to be influenced by two causes, the one a tendency to precocity and premature development, the other the natural law by which all organisms tend to reproduce the germs of their species before they themselves decay. Sexual precocity is a certain evidence of a tendency to early nervous degeneration, and this is a quality which has existed in the being from the time it was laid down in the germ.

Of the various causes operating to produce a

potentiality of insanity in succeeding generations, perhaps one of the most powerful is to be found in marriages amongst the tainted, more particularly when such marriages are between blood relations.

In our last lecture we discussed the subject of consanguineous marriages, so that we need not enter upon it again now; but since selection on health considerations is practically impossible in questions of marriage, consanguineous marriages ought to be very strongly condemned.

In the present condition of English Society, however, such marriages cannot be prevented, neither can we interfere in marriages among tainted persons; it is, too, a matter not yet ripe for legislation, and until society generally becomes strongly impressed with the wrongfulness of these unions, there will not be any likelihood of legislative interference.

You will, often, be asked for your opinion as to the propriety of marriages in particular cases. But I think it no part of a medical man's duty to do more than to give his advice on such matters in broad generalities, for the persons most concerned have usually made up their minds before asking anyone else's opinion.

We have seen that the union of blood relations, when unselected, tends to deterioration in the issue; we are bound to state this when asked, but we are not called upon to bear the opprobrium of interference.

Another case in which we should express a strong opinion when asked, is that in regard to the marriage of those who have been lunatics, or who are, or who have been epileptic. That such marriages are highly injudicious no one can have a doubt; and that the issue of such marriages, tends to perpetuate diseased conditions, which might be, in a great degree, if not entirely, stamped out, there is very little question. But whatever may be the opinion of the wise and learned on the subject, I cannot but sympathise with those who aver that it would be a cruel interference to impose any legal restrictions upon such subjects.

I will ask you, nevertheless, to note, that the potentiality of insanity like all other constitutional defects, has a tendency to die out. It will terminate by producing sterility and death in the worst subjects, and in the best, it will follow the natural law of decline—"The beneficent law," as observed by Sir W. Gull in his eloquent address at Oxford, *On Medicine in Modern Times*, "to which we owe the maintenance of the form and beauty of our race, in the presence of so much that tends to spoil and degrade it."

It is not a flight of imagination to suppose, that were it possible to make better selection in marriages, that insanity might be reduced to a minimum, for as Sir W. Gull remarked, in the address just spoken of: "The effects of disease may be for a third and fourth generation, but the laws of health are for a thousand." Under better circumstances, and by reason of the tendency to recovery after the sterile degeneracy had worked its work, and the worst instance had died out, the potentiality of insanity in a family might die out too, were it not nursed and nurtured by injudicious marriages, and by excesses and excitements of many and various kinds.

This subject is highly important, for whilst with our advancing knowledge, the percentage and deaths from phthisis, fever, and general diseases are constantly diminishing, deaths from brain and nervous diseases are constantly increasing.

It is certain that insanity may succeed to epilepsy; and, as the conditions which set up epilepsy are transmissible, it is but a natural step for such subjects to transmit the potentiality to the next generation.

Idiopathic epilepsy is associated in the same train of neuroses as that to which insanity belongs. But traumatic epilepsy will readily drift into insanity, and the subjects of it may progenerate a race predisposed to insanity and epilepsy.

I was lately asked the question in reference to an infant, whose father became the subject of traumatic epilepsy shortly after his marriage, whether the child was likely to have inherited the epilepsy of its father. The father's epilepsy was very severe, but from the history, the child must have been conceived before the accident, which set up the epilepsy in the parent occurred. The father's former health and history was good, and the mother's history was good also. The child has never had convulsions, is perfectly healthy, and there is no reason to suppose that it will ever

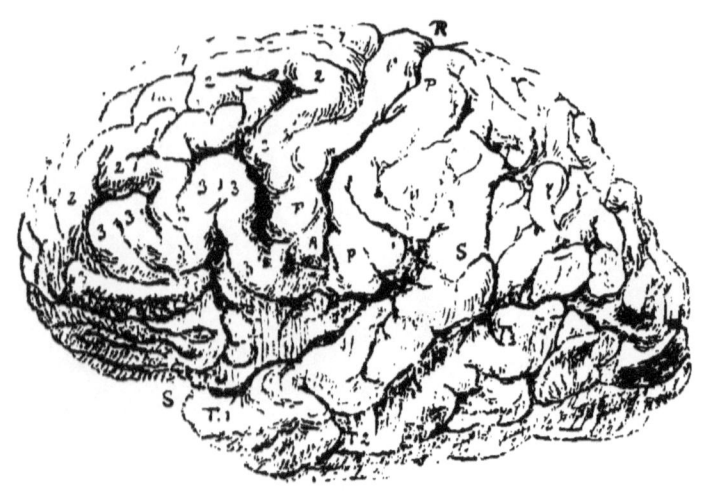

Figs. 1, 2, 3. First, second, and third frontal convolutions.
R.R. Fissure of Rolando. S.S. Fissure of Sylvius.
I. Island of Reil. T 1, T 2, First and second temporo-sphenoidal convolutions.
F.F. Transverse frontal convolutions. P.P. Transverse parietal convolutions.

suffer from nervous disease, though future children of this man, should procreate any, may be epileptic or idiotic, or may become insane. Such a question might be of great importance in proposals for life insurance, as also to the individual in his family and personal interests.

In reference to transmissibility of potentiality, the subject of syphilis should be considered.

That syphilis may powerfully affect the nervous system, and thereby become a primary cause of insanity is certain, and that it also originates a predisposition there can be but little doubt. Numerous cases of syphiloma, pressing upon and destroying the surface of the brain, and setting up epilepsy have been recorded, and the sufferers from epilepsy may at any time exhibit mental symptoms. Changes in the calibre of the cerebral vessels, too, from thickening of their lining membrane and deposition of areolar tissue around them, may give rise to atrophy of the brain, in consequence of which, the phenomena of morbid mind appear. The morbid conditions of brain seem to be transmissible; but we are in want of more facts. I have very little doubt, that numerous cases of insanity arise from syphilis in parents, but, as we saw in considering the etiology of idiocy, direct evidence is not often obtainable. I had under my care one case, in which none but a syphilitic history could be traced. The patient, a girl, was the eldest of six children, one of whom had notched syphilitic teeth, another had screw driver teeth, and a brother, the fourth child, was semi-idiotic. There was no trace of any history of insanity on the side of either the father or the mother, but the father confessed to having suffered from syphilis before his marriage. Such facts must of course only be taken as grains in the balance, but they help us to look for a possible origin of the potentiality in obscure cases.

I have had lately under my care, four instances of hemiplegia traceable only to syphilis, and Dr. Hughlings Jackson has published[*] numerous other cases, and he has very ably elucidated the subject of syphilitic diseases of nervous tissues.

I must now say a few words in reference generally to an insufficient supply of nourishment to the brain.

Starvation as a cause of insanity, from imperfect nourishment, is evident, and I have already brought forward numerous facts to show prominently its influence in the production of insanity.

A very remarkable circumstance to be noted in considering starvation in relation to cause, is that the greatest proportion of the late increase of insanity has been observed in the poorer agricultural districts, where low wages and expensive provisions necessarily tend to ill-nutrition and impoverishment of health. Under the depression of semi-starvation any defect will show itself, and, as we observed in our consideration of mania, mental aberration readily appears in the predisposed when they are

[*] Cases of diseases of the nervous system in patients, the subjects of inherited syphilis, reprinted with slight alterations from the *Transactions of the St. Andrew's Graduates Association*, Vol. I, 1868, and *London Hospital Reports*.

exhausted by want of food. How far starvation may operate as a primary or predisposing cause cannot, from our present statistics, be judged.

It will be useful, however, to mention a few facts illustrating unequivocally the influence of the circulation in the production of variation in the cerebral function. I have recorded several cases from my own observation; and others, most interesting in their details, were mentioned by Dr. Burrows,* in his Lectures delivered before the College of Physicians, in which mental symptoms followed upon morbus cordis. Dr. Burrows further illustrates his point with a case from the writings of Dr. Abercrombie,† a case showing how posture alone was able to influence the sense of hearing. The patient when standing or sitting upright was very deaf, but when he lay horizontally, with his head very low, he heard perfectly. If, when standing, he stooped forward, so as to produce flushing of his face, his hearing was perfect, and upon raising himself again he continued to hear distinctly as long as the flushing continued: as this went off the deafness returned.

Dr. Burrows expressed his opinion that the deafness and the restoration of the hearing was caused by the varying amount of vascular pressure. It appears to me that it was rather due to diminished supply, the amount of blood circulating in the head in the erect position being insufficient to main-

* *Op. Cit.*
† *Pathological and Practical Researches on Diseases of the Brain and Spinal Cord.*

tain the cerebral function perfectly and in its integrity.

Another case mentioned by Dr. Burrows, as recorded by Dr. Andrew Combe,* was that of a boy, eleven years and a half old, with a languid circulation, and feebly-developed chest and heart, who seemed to be two different characters when sitting up and lying down. In the former attitude, when the head was scantily supplied with blood, he looked apathetic and was sullen; whereas, upon lying down, when the circulation was assisted by the force of gravity, his real powers became manifest, and he was animated, talkative, and highly intelligent.

A third case, also mentioned by Dr. Burrows, was recorded by M. Bricheteau in an essay on the influence of the circulation on the cerebral functions.† It was that of a student whose memory was treacherous, and who was in the habit of placing himself, not merely in the horizontal posture, but with his lower extremities elevated, and his head depending, when he wished to study with most effect. Dr. Burrows appears to have considered that the mental activity was directly as the momentum of blood in the cerebral arteries, and Bichat, quoted by Dr. Burrows, seems to have been of the same opinion. Though more or less the fact, the actual fact is rather that the mental activity is directly as the nutrition, or directly pro-

* *Physiology*: 1841 10th Edit.
† *Journ. Complement des Sciences Medicales*, Vol. iv. 1819, Paris.

portionate to the nutrient supply, for although the cases quoted prove the proposition of Dr. Burrows in regard to the momentum, they also confirm the proposition in respect of the nutrient supply, whilst numerous instances might be cited showing that whatever the momentum, the cerebral function will be interfered with if the quality of the circulating fluid be such that it does not perfectly nourish. Cases of blood poisoning, and also of spanæmia, furnish examples.

That some subjects whose brains are imperfectly nourished from bad circulation, or from bad or poisoned blood, become insane, is certain, as we have already seen. It has yet to be determined in how far these causes are predisposing, or whether they are only exciting. It has also to be determined to what extent the influence of these causes is capable of transmission. It is not uncommon to find abnormal cerebral phenomena in children in whom hereditary predisposition to nervous disease is denied, but in whom sub-acute maniacal symptoms, garrulous or loquacious gabble, or melancholic depression, associated with homicidal or suicidal tendencies sometimes appear. You will always find that such children are out of health, many of them are the descendents of gouty, rheumatic, or tuberculous parents, and others of parents who suffer from constant emotional disturbance, or, as I have elsewhere described it, of sympathetic depression. In these children deterioration of all the tissues is constantly present, and whenever such a child falls short of its normal nutrition, the brain is the first organ to show the want, because the brain in childhood is the most active, or rather the most rapidly-developing organ, and is in consequence very susceptible of slight deteriorating influences. The highly excitable state of many children is very instructive, particularly those with a tuberculous history. They are often very precocious; for a time they are bright, and almost intellectual, but invariably they are irritable and wayward. Often they develope epilepsy, usually in the form of *petit mal*, which is frequently overlooked; they constantly have hot heads, and complain of headache, and frequently, like window-grown and over-drawn plants, they die before any of their tissues become fully developed.

I should mention that diminution in the calibre of the cerebral vessels produces different defects, as the change is general or local; when it is general, it usually produces dementia; when it is local, it may produce a great variety of paralytic symptoms, sometimes amounting to hemiplegia. A not unfrequent defect is loss of memory; a more striking result is partial or complete aphasia, which may be regarded as allied to mental imperfection, and is often more or less associated with it. The patient first seems to forget the word, and then uses a wrong word, and lastly, retains only two or three words which are used for everything. These cases are not very common, but the *post-mortem* examination of them invariably discloses wasting of certain convolutions.

When the defect appears as aphasia more or

less complete, it is often followed by, in fact the aphasia may be looked upon as a percursor of, hæmorrhage from the middle cerebral artery or its branches. Temporary aphasia, indeed, is not uncommon, but in many of the recorded cases the patients have died of cerebral hæmorrhage. I noticed a very characteristic case of aphasia in Stephen Ward, some months ago, under the care of Dr. Moxon; the patient had suffered from syphilis, and had a large tumour connected with the bone in the left temporal region, and it is not improbable that, either the thickening of the bone extended internally, and pressed upon or arrested the blood supply to the convolutions in which the memory of words is stored, or that the middle cerebral artery was altered in its calibre; these, are at least, possible explanations of the aphasia.

I have seen one brain in which the Island of Reil and the surrounding convolutions were wasted, and the patient suffered from aphasia, but I did not see the patient.

I cannot enter fully at this time upon the phenomena of language. The subject of aphasia has been studied very carefully by Dr. Hughlings Jackson;[*] and also by Dr. Bateman of Norwich, who has written an excellent work on the subject.[†] It has not appeared consistent with Dr. Bateman's

[*] Paper on the *Physiology of Language*, read before the British Medical Association at Norwich; and other papers.

[†] *On Aphasia or Loss of Speech.* By Frederick Bateman, M.D.

own observations, for him to endorse the opinions of M. Broca, whose views most of you are probably acquainted with. But Dr. Bateman has collected many valuable facts in connection with the subject, and you should peruse his book. My own observation has led me to look for a local seat of articulate language, and it appears to me that if Broca's convolution be not the only seat, it is, nevertheless, a very important seat, the neighbouring convolutions perhaps sharing in some degree the duties of Broca's convolution.

The drawing upon the wall is enlarged from the engraving, which forms the frontispiece of Dr. Bateman's work on Aphasia, and it was printed from a block sent by Professor Broca to Dr. Bateman. I have made the sketch in order to show you Broca's convolution and the region of language. The convolutions marked 1, 2, and 3, respectively, represent the first, second, and third frontal convolutions, the latter being the one rendered famous with Broca's name; it is the posterior portion of this which Broca has demonstrated to be more particularly concerned in giving a seat to the faculty of speech. It is probable that language is intimately associated with impressions of motion, and those impressions are laid down in the centres which govern the motions of the organs of articulation. A child learning to speak, repeats a word over and over again till he utters it correctly; the articulation of the word becomes easy by practice; the impression of the word grows, in the governing centre, with the constant repetition, and we can scarcely doubt the fact that governance over all motions resides in the convolutions. The fact, that destruction of the posterior portion of the third frontal convolution, causes a cessation of the power of articulation and an absolute loss of words, is a strong confirmation of this view, and it localizes languages in that spot.

This view is, I think, supported by many of the observations made by Dr. Hughlings Jackson, as well as many made by myself upon local epilepsies in association with circumscribed lesions of the surface of the brain ; these clearly indicate, not only that the governance of motions resides in the convolutions, but also that the function of governance, over the various parts of the body, is not distributed diffusely in the convolutions, but centred in definite local spots or regions.*

To complete our survey of the physical exciting causes we must enumerate those which tend to general exhaustion, as seen in the simple cases of the so-called puerperal insanity, of which we have had some cases under consideration. On the subject of insanity associated with pregnancy, childbed, and lactation, I need hardly add to what I have already said, neither need I detain you longer upon the subject of wasting and exhaustive dis-

---

* Whilst these pages were passing through the press, the publication of the experiments of Professor Ferrier in the *West Riding Asylum Reports* for 1873, has added a valuable confirmation of the deductions from observation, upon local seats of governance or control in the brain.

ease, as fever, tubercle, cancer, and uterine affection.

Chloroform as a cause ought to be noticed; I have seen one case of dementia following its administration; and a subject which should be mentioned in regard to cause is railway travelling. It has been the experience of numbers of people who have travelled twice daily for any length of time, between Brighton and London by express trains, that they were obliged to relinquish the daily journeys; they having suffered from symptoms of nervous exhaustion in consequence. The effects of railway travelling in the predisposed should be thrown into the scales on the side of exciting causes.

Statistics have not yet shown railway travelling to be a primary cause, although, since the day when rapid locomotion was inaugurated, insanity has been rampant and shown itself as a very common disorder. It appears highly probable that the motion of railway travelling tends to exhaust the brain by the vibrations conveyed to it through the spinal cord, and this exhaustion, it is supposed, may prove sufficient sometimes to upset the balance of a mind which is trembling, upon the knife edge as it were, between sanity and insanity.

I have already mentioned a case in which the acute attack first made its appearance at the end of a railway journey, and I have several times observed patients suffering from insanity become excessively excited by railway travelling. Excesses in walking, too, are most pernicious; the subject of incipient insanity often seeks relief in inordinate exercise, but the effect of it is only to increase the tendency to exhaustion. The wear and tear of daily life in these high-pressure times, seems to produce changes in the lining of arteries of the brain, and as a consequence, waste of brain tissue and premature senility is not uncommon. But these effects are not constant. It often appears that the mental powers improve after middle age, even though the intellect be very greatly exercised. This observation therefore, is consistent with the view of potentiality, that excessive brain-work only induces insanity in the predisposed, though excessive brain labour gets the credit, popularly, of setting the morbid change agoing. When, however, the emotions are involved in the motives for the excess of brain work, a more powerful cause comes into operation. According to Morel: "Great intellectual activity when unaccompanied by emotion does not become the excitant of insanity," and he says, "It is when the mind is the theatre of great passions that it is moved."

The case of an artist I mentioned to you may be taken as an illustration. But in his history we found a predisposition, and it is hardly possible that his mind would have lost its balance had he not been predisposed to brain disease.

Moral Causes of Insanity.

The moral causes of insanity, as they are called, require a brief notice here, since they impose an undue tax upon the brain and thereby become excitants. But they do this principally through the medium of the emotions, and I shall have to allude to them again under that head.

Emotion, or the excited manifestation of one of the emotions, is often a very prominent feature in insanity; but emotion is only one of the constituents of mind, and its disturbance is not necessary for the production of insanity.

Disappointment, long pent-up anxieties, love, surprise, fear, apprehension, expectation, are all frequently assigned causes of insanity, but their influence would, probably, be unimportant in subjects in whom no predisposition existed.

You will hardly go into an asylum in which you will not find one, or perhaps many cases, whose history is that of a love-story, blighted affection or broken heart as it is often described; but the love disappointment is only the breath that fanned the insanity into flame, which might as readily have been kindled by the excitement connected with marriage.

Surprise is often found to be the excitant of mental derangement. I saw one case in which a man became insane upon the sudden accession to £100. Dr. Winslow mentions a case in which a gentleman realised £60,000 in a fortunate speculation, and immediately became demented.

The pain of stricken conscience is often the excitant of insanity. A child will tell a lie, and shortly after show signs of mental aberration; or a person will take an oath, which will become the dominant idea excluding all, or almost all others, from the mind.

In all such cases, you will find that the patient at the time of the invasion of the attack is out of health; many of them recover, but the prognosis is doubtful if they are cachectic, debilitated, and weak.

But one of the greatest moral exciting causes is to be found in the artificial life which we lead in the present day. Many of its influences conduce to insanity, and not the least of these is the unnatural condition of continence, which prudential considerations enforce upon a great number, particularly upon women. Apart from the fact that the percentage, in the number of women over men, in this country is nearly three, a large number of men will not marry from prudential reasons, and thus a large percentage of women never have the chance of marrying and fulfilling the reproductive act. The desire for fulfilling the act of reproduction is strongly implanted in women, and the non-satisfaction of it is reflected back upon their nervous systems; and though statistics showing this as a cause of insanity have not, yet, been furnished with regard to the question, nevertheless, the cases which come before the physician strongly tend to prove the truth of his *a priori* conclusion on this point.

The operation of the cause is even more powerful now than it was two or three generations back. In the present age a premature development is cultivated. Girls are forced into positions of womanhood at tender ages, and taught to regard marriage as the one aim of life.

> "Motus doceri gaudet Ionicos
> Matura virgo et fingitur artibus;
> Jam nunc et incestos amores
> De tenero meditatur ungui."*

> "The ripening virgin joys to learn,
> In Ionic dance to turn,
> And bend with plastic limb;
> Still but a child, with evil gleams
> Incestuous love's unhallowed dream,
> Before her fancy swims."†

Is it then a wonder if disappointment on this head adds its weight to the deterioration resulting from the unaccomplished design of nature? The necessity for the performance of the sexual act is one of the most constant of natural laws, running throughout animals and plants. Nature will not be deceived. If you rob her in one place you must pay her back in another. And if you deprive an organization of one of its most important functions you must expect to see deterioration as the result of the deprivation.

The girl of the period, cultivated to early sexual maturity, and with an education fitting her only for frivolities, as though dress and enjoyment were the

---

* Horace *Odes*, Book III, Ode vi.
† Translation by Theodore Martin.

aim and end of existence, and encouraging in her, selfish impulses and sensibilities, is early disappointed if she does not marry; she then feels a craving for something which she, perhaps, cannot explain; as a rule she seeks for satisfaction in anything that is attractive, and too often, contemplation and brooding over her supposed miseries suggest devotion and religious exercise, as the nepenthe to soothe her morbid longing. But ill-timed, and ill-judged religious exercises soon develope excitement; this perhaps soon subsides in the more healthy and vigorous; but if the subjects are predisposed to insanity, the religious excitement soon becomes fanaticism.

Religious excitement is a very popular explanation of the cause of insanity, but the supposition is perhaps even more groundless than that of many popular beliefs. The religious fanatic, however, furnishes an excellent example of a mind incapable of reasoning, and the fact that such a being should develope some definite form of mental aberration is not by any means surprising.

Society has some grave charges to answer, in regard to its influence upon the present and the rising generation. The habits of luxury inculcated by the growing wealth of the country among the individuals of one class, and the stinging poverty of untold numbers of another class, tend to nurture the predisposition which might under neutralizing inaction, or more equable distribution, have died out.

The combined influence of luxury and poverty

is best seen among the class which is really poor. Not the artisan class, who are comparatively rich; nor the class of whom we have statistical information, called paupers, but of a class far worse off. I meam the middle class—a class educated up to the luxury of the times, but forced to struggle with want and privation. The class educated to the standard of the rich, and equal, or often superior to them in intellectual attainment, but poorer than the artisan, because the only labour the members of it are capable of rendering is the labour of the head, not of the hands.

In the present age head work is counted as of little worth, and the dependent upon it is goaded on into utter drudgery, in order to gain his insufficient meal.

There are hundreds of merchant princes in this great city who amass wealth by grinding half-starved clerks at a desk from morning until night, men who live in superabundance and luxury, who pay for one dinner a sum that will pay a clerk, or perhaps two or three, for a week, and yet who would discharge from their service, and consign to beggary, any one of these clerks who asked for the crusts that are daily carried from the table to the dust heap.

I know that there are alienists who argue that with increasing civilization insanity decreases, but statistics show mental diseases to be increasing, at all events in this country, and we cannot but recognize that the influence of our social systems, and of the age we live in, account for the increase in a great degree.

Society at the present time, deserves to smart under a severe lash, and one as bitter as, or more bitter than, that addressed to the Roman people in their age of luxury, by Horace, in the ode from which I have quoted, and which was strongly prophetic of the Roman degeneracy.

General Note on the Pathological Anatomy of Insanity.

On the general pathological anatomy of insanity there are yet one or two negative points upon which it is right that I should lay some stress. The evidences furnished by a pathological anatomy of insanity are atony which we cannot see, and atrophy and effusion which we can see; we also find changes in the vessels, and excess of connective tissue, and thickening and ossification of membranes. Sometimes in the insane we find softening, and much has been made of it. It is highly probable, however, that many of the pathological conditions of the brain, found on the post-mortem table in autopsies of the insane, are accidental, and particularly the condition of softening. I know of no form of insanity, for instance, which can be said to be the result of "softening of the brain." I have often heard people speak of "softening of the brain" in regard to insanity, but I candidly confess I do not know what is meant by the expression, and I have never yet been able to find anyone who could explain to me what he meant by it. Red softening occurs from passive congestion, which follows an arrest of the blood stream. Red softening, too, may result from active congestion, and appears in circumscribed patches which would speedily go on to abscess, as may be seen in acute inflammation, or in cases of pyæmia, but the condition is usually secondary, and rarely produces symptoms of insanity. Yellow softening occurs from the cutting off of the nutrient supply, as by plugging one of the arteries of the brain with an embolus; associated with this you sometimes see loss of memory and more or less fatuity; and white softening is the result of œdema; but none of these states give rise to a special form of insanity, though any of them may be, and sometimes are, found in the *post-mortem* examinations of insane subjects. That which is popularly called softening of the brain, is I believe paralytic dementia, but we have seen that in paralytic dementia the condition of the brain is one of atrophy, and not of softening.

GENERAL NOTE ON THE TREATMENT OF INSANITY.

The general treatment of insanity requires a short comment. It may be summed up in the words, rest and nourishment. Patients should be removed from, or relieved of exciting causes, and every attempt should made to nourish their brains, and to restore their general bodily health and vigour.

Change of scene is of immense value; but regard must be had to the fact that change of scene is often connected with exhausting travelling, and the latter may do more harm than the former good, so that very great caution is necessary, and whenever change is made, the immediate consideration should be to provide rest as soon as possible. In acute mental disease railway travelling should, above all things, be avoided. A railway journey cannot be taken without some excitement and a considerable waste of nervous energy, highly prejudicial to the patient and often irrecoverable.

Passive states, however, require the stimulus of of excitement. A gentle stimulus may set in motion the machinery of a mind that has ceased to act, and which without a reviving stimulation would have sunken into absolute and complete dementia. The stimulation, however, must not be too great, or too long continued; it may consist of any available entertaining source of amusement in which the patient can be induced to take an interest, let it be music or gardening, or needlework, or reading, or drawing, or some art or manufactures, or even card-playing or dancing. If, in passive states, an interest can only be developed, be that interest so small even as the feeding of a bird, or the culture and growth of a plant, you may set in motion a new and active phase of mind. If you can move the emotions into new states, there is some hope for the patient, but, failing to rouse a new interest, I fear that the case would have to be pronounced hopeless.

Bodily exercise is often of great use when acute

symptoms have altogether subsided, but during their continuance, rest in bed should be strictly enjoined. When, however, the case is free from the active conditions which waste the nervous energy, the patient may rapidly improve under the stimulus of exercise, which should consist of either ordinary walking, riding, or gymnastics. Bodily exercise tends to promote secretion and excretion, and acting as a slight stimulant to the vital powers, tends to place a chronic case in the most favourable condition for recovery.

There is, however, one question in regard to the general treatment to which you must give due consideration, viz., whether asylum treatment is necessary or not; and apart from the injury that may be inflicted upon family interests by the declaration of insanity, the physician is bound to give some consideration to popular feeling. I would maintain that it is a physician's duty to correct popular errors when they militate against the good of the patient, but popular feeling is not always wrong, and as regards lunatic asylums it is often right.

The consignment of a patient to a lunatic asylum is, in the popular idea, to brand him as a person to be regarded with curiosity, at least, if not caution or suspicion, and his family is always afterwards regarded with some degree of doubt, if not also branded as crazy; and what is more is, that the interests of two or three generations may be needlessly affected by it.

The interests of families are bound up and

cemented together by their marriage relations, and the stigma of insanity is, and not perhaps unjustly, so great, that untold sacrifices,—not only of money and property, but of feeling and affection, and the nobler possessions of men,—are, not unfrequently, made on account of it.

Over and over again it happens, that a case of insanity arises in a family from some extraneous cause, as a blow, or a sun-stroke, but it would be the height of cruelty to brand such a family with the stigma of insanity. Cases occur in which mental symptoms complicate the last stages of many somatic diseases, particularly, such for instance as many forms of paralysis, fevers, and exhaustive diseases generally; or a patient may suffer, as we have seen, from progressive paralysis, such paralysis having occurred in his family in his case only, his mind with his body having become enfeebled, the alienists say that he is suffering from general paralysis of the insane: but feebleness of mind does not constitute insanity, and it would be hard indeed to call such patients insane. If there be no other mental symtoms than mental feebleness, surely it would be cruel to the utmost degree to brand the family, in which such a case occurred, as predisposed to insanity, and thus ruthlessly and without warrant destroy the life prospects of the junior members, which the consignment of the unfortunate paralytic to an asylum surely would do.

As a matter of principle, I should strongly recommend that a patient never be sent to an asylum, if such a course can be avoided. There is no law prohibiting the treatment of a patient at home. The lunatic is not a criminal to be put under locks and bonds; and it is only when he disturbs the public peace, or when, by cruel and illegal treatment, other people infringing the law as regards him, that authority can interfere on his behalf.

A man may have a delusion, or a dozen delusions, of a harmless character. Is he in consequence to be made a sort of state prisoner and alienated from his home? Not long since, I saw a gentleman who had a number of delusions, and he had had them for years. He lived with his aunt and was submissive to her in everything, and at the time I saw him, it seemed that the only delusion of any consequence, was one which might have led him, at any time, to make a disturbance. It was, therefore, necessary for his own safety that he should not be allowed to go out unaccompanied, but it would have been the height of injustice to deprive him of his liberty; and yet so extravagant and absurd were some of his ideas, that the merest tyro in mental disorders would have discovered his insanity at once. Such patients are often aware of their insanity, and though they may profess to object to have someone at all times with them, yet they usually assent if the servant or attendant is kind. Insanity should be considered as a state in which the individual is out of health, and one in which it is necessary, in consequence of the resulting feebleness, to have the patient constantly attended, and no law can oppose this system.

But removal from home is sometimes necessary; home scenes may be distasteful to the patient, and become sources of irritation; removal is then one of the essentials of treatment; or the patient becomes violent, and the accommodation of his home is such that a risk of danger, either to himself or others, is run by retaining him; it is then necessary to remove him; or the patient may be noisy and disturb and become unbearable in the household or the neighbourhood; removal in such a case is necessary; or the patient is wet and dirty, and the accommodation of home is insufficient for his requirements, or his means are inadequate to support the great expense which a lunatic in his own home entails. It then becomes necessary to remove him.

Sometimes, too, the moral influence of an asylum is useful and beneficial. Patients commencing to recover are sometimes stimulated to emulate other convalescents in an asylum, and after comparing their state with that of other inmates, begin to exercise the faculties of judgment or reason, which, for want of the stimulus, might have remained long in abeyance.

I remember reading of three patients, each of whom believed himself to be the Holy Ghost, being brought together. One of them it appears began to reason that if either of the others were the Holy Ghost, he could not be. The other two seem to have remained unshaken in their belief. There can, however, be no doubt, that the brain of the patient who commenced to reason on the subject was regaining its tone, and that the patient himself was on the high road to recovery. Not only, however, for the occasional value of the comparison, but for the regime, an asylum is of great use; and above other things, the accommodation for baths and other hygienic measures, which in every asylum ought to be a first consideration, provides an essential part of the treatment, which at home is often a matter of difficulty, if not of impossibility; in addition, too, the expense of asylum treatment is less than private treatment, and this is often a matter of the greatest importance to the patient or his friends.

The hygienic conditions of many houses would tend to confirm insanity in some cases. I mentioned a striking instance, in speaking of melancholia, and of the depressing effects of living in a cheerless lodging.

The subject requires early consideration. Removal from cheerless and depressing influences is absolutely essential. As a rule, a lodging house is the very worst place an insane patient can be in. When it is possible to obtain treatment in the house of a skilful medical practitioner, who will devote time, care, and attention to the case, (provided the hygienic arrangements of the medical practitioner's house are suitable), the patient so placed will be under good influences for recovery. Such a course of treatment is necessarily expensive. It cannot be undertaken, to do justice to the patient, for a small sum, but if the means of the patient can provide it, it is, next to home treatment, the most desirable.

For chronic cases, no treatment can be more humane than that of the reception of single patients into private families, or the principle of the boarding-out system as practised with so much success in Scotland, or the system adopted at Gheel in Belgium, where the lunatics are gathered together in a village, but allowed to wander unrestrained within certain ranges, and to amuse themselves in and about the country. The success of this system is highly worthy of consideration and trial in England, if not of a wide adoption.

LECTURE XII.

Morbid States of Mind.

General symptoms of Morbid states of Mind—Delusion—Illusion—Hallucination—Delirium and Delusion—Classification of Insanity—Enumeration of Phenomena of Morbid States of Mind from Insanity of Sense—Insanity involving the Emotions—Sympathetic Depression—Moral Sense and Moral Insanity—Conclusion.

Delusion is considered to be the only conclusive evidence of insanity, but illusion and hallucination are very often indications of morbid mental states. Insanity without apparent delusion is also not uncommon, so that we must consider the general evidence of morbid states of mind somewhat closely.

The terms—illusion, hallucination and delusion, were tersely defined by Jamieson, in his excellent lectures published in the *London Medical Gazette* in 1850, and he remarks that they are terms "of very common occurrence in medical writings," but, "not always used very categorically." "An illusion" he says, "is a mistake of the senses, a false perception; a hallucination is a baseless creation of the fancy; a delusion is an illusion or a hallucination misleading the judgment and governing the conduct."

We often hear people speak of errors in observation, and imperfect experiences as delusions; in such instances, however, the term is misused, and expresses what is not fact. For if every erroneous conclusion is to be considered a delusion, then delusion must be divided into two classes, sane and insane. But such a division cannot be admitted. Errors in judgment arise from imperfect knowledge and bad logic, and the premises may be imperfect though not false; the reasoning may be bad, but this does not imply disease, whereas in delusion the premises are false, though there may be no fault in the logic.

An illustration is to be found in the common experience of an ever-ready kindliness, which is always willing to regard and trust every acquaintance as a friend. The confidence is often misplaced; but the belief in the friendship cannot be considered a delusion of the victim, it is but an error of judgment arising from imperfect knowledge. A further illustration is seen in the too common instance, in which one person under the belief of the honour of another, becomes surety for his good faith but is deceived; this again is a simple error of judgment arising from bad logic, not from disease, whereas should a person thrust his hand into the fire under the belief that the fire will not burn him, he must be labouring under so much disease that past experiences cannot be aroused to correct the present impression; in other words his judgment has been misled by a mistake of his sense, or by a baseless conception of his mind, and he is labouring under a delusion.

Illusion is an evidence of mental defect, but it is to be measured by degree.

The conjurer or the ventriloquist calls his deception illusion, but it is not so; were your eyes or your ears educated to follow the rapidity of motion or sound upon which his deception depends, they would not be deceived at all; the case is different, however, when ordinary sight or hearing makes mistakes. In simple illustration, if on leaving a jovial gathering you mistake a gate post for a giant, or a white cow for a ghost, it will be evident that either from the toxic effects of alcohol, or some allied cause, your brain was not doing its work perfectly, and that all the currents which should come into play to correct visual impressions, are either not in motion or only imperfectly doing their duty. This may be further and more forcibly illustrated by consideration of the illusion of a phantom ship, which has over and over again been detailed as occurring to the famished and exhausted sailor, and most graphically described by Coleridge in his *Ancient Mariner*, or by the illusion of cataracts and the sound of rushing waters, as experienced by thirsty travellers in the desert who are so reduced as to be unable to correct, by the impressions of former experiences, what they see or hear, and they imagine that whatever reaches their sight or sound must be the long hoped-for relief of which they stand so much in need. Whatever observers may choose to call this state, it is a loss of the faculty of comparison, temporary, perhaps, but nevertheless a loss, and in consequence, the individual cannot be considered for the time being as of sound mind.

Hallucination, probably, differs pathologically from illusion in that it seems to be dependent upon a greater degree of brain imperfection.

But hallucination, or baseless conceptions, resembling impressions on organs of sense, have their origin in a brain so imperfect in its action that no attempt whatever is made to correct the errant currents of thought, which may, like a dream, be set in train by a chance cause.

Dreaming, in fact, is a very fair illustration of hallucination; a current of thought gets into motion whilst the brain is in a greater or less degree asleep, and the faculty of comparison and judgment more or less, or altogether in abeyance; and the reproduction of impressions which have at various times been formed in the brain, give rise to those extraordinary mental panoramas which are called dreams, and which are looked upon by the vulgar as supernatural. Dreams, indeed, were formerly considered as revelations and inspirations, and the medium adopted by the Almighty to make his will known to men.

There used to be many, and there are still some firm believers in the Divine origin of dreams, and it will, no doubt, be a long time ere this superstition is effaced from the popular mind. As another illustration of hallucination, I will recall to your minds a state I referred to in a former lecture, viz., that of an overwrought brain, such as that of a student, who has over-read himself, and retires to rest, unable to suspend his consciousness sufficiently to fall properly asleep: he lies in a dreamy half-conscious state, revolving before him the subject of his work, in an exaggerated but imperfect impression, which he cannot clearly define or correct. This state, which in its mechanism is exactly like dreaming, is vivid and painful, because of the exhaustion of its victim at the time of its occurrence, and it is a state which if not speedily recovered from, would eventuate in insanity.

The most important forms of hallucination, however, are those in which the phenomena occur whilst the subject is fully awake and about his daily calling.

To see a ghost, or to hear a voice which has not a foundation or an origin in some external physical fact, is an evidence of some internal disorganization or defect; the man's brain in whom it occurs may be compared to a barrel organ when out of order, and in which a discordant jumble of notes makes an apology for music.

A gentleman told me one day that, whilst sitting in a cave at the sea-side, he saw "the Spirit of God moving on the face of the waters," and after discussing the subject with him, he confidently assured me that he believed in the reality of his vision. I expressed to his friends my belief that he had been the victim of hallucination, which was fast becoming delusion, and within a fortnight it became necessary to confine him in an asylum.

Effect without cause cannot occur, and except from the imperfect motions of a disorganised brain, no such baseless creations can form themselves into permanent impressions.

I have met with several instances in which the patients have assured me that they heard the angels singing, or that they had heard God speaking to them, or have seen visions of God or angels in their room; all these, and all similar cases, are examples of imperfect mental action, and a loss of the faculty of correction or the power of judgment. In all such cases, it behoves us to watch closely and narrowly for other symptoms; and it consists with my experience, that unless the patient speedily regains the power of correcting hallucination by reason, that he rapidly developes such other symptoms as unmistakeably announces him to be insane. If a person becomes the subject of hallucination, and can correct the impression by an effort of reason, he may be considered as sane. But if he cannot correct the impression, he cannot be of sound mind; and if he acts upon a belief in, or obeys the promptings of his hallucination, he must be mad.

Hallucination may be temporary, and may arise from causes, such as exhaustion, alcohol, fever, or other poison, the effects from which are speedily recoverable. But when the cause is not speedily removable, and especially if there be a history of insanity or nervous disorder, such as epilepsy, catalepsy, sleep-talking, sleep-walking, or other similar abnormality, the case ought to be looked upon as one of gravity, and one in which some severer symptoms are likely to be speedily developed.

The most common forms of hallucination are

those of sight and sound. Sometimes they are of the various subdivisions of the feelings. Sometimes they are of taste, sometimes of smell; and it is a curious fact, that in some cases where such hallucinations are the accompaniment of chronic or confirmed insanity, the patient has lost altogether the sense which is the subject of his hallucination.

I related to you the history of a patient whose case was recorded by Griesinger, who suffered from hallucination of smell, but in whom it was afterwards discovered that the sense of smell was altogether wanting.

The same author mentions another case, in which the patient used to lick the wall, whilst labouring under the belief that he was sucking beautiful fruit.

Either illusion or hallucination may become delusion, when they are easily recognized; the judgment is then misled or warped, and the conduct of the person is usually singular and altered.

On visiting a patient one day I found him with a number of pieces of straw in his hair, and on asking what they were there for he said that they were angels, and he refused to allow them to be touched. Another patient I used frequently to see, was in the habit of putting his ear to the tables and chairs declaring that he heard sounds issuing from them; these are instances of delusion arising from illusion of sight and sound which misled or warped the judgment and governed the conduct.

Delusion from hallucination, however, is more common. The patient, I mentioned in a former lecture, who stated that "God came to him in the night and told him to cut his throat" is a remarkable example.

A few years ago a woman was tried for murdering her children by throwing them out of the window, but acquitted on the ground of insanity, as it was shown, that she got up from her bed and committed the act whilst labouring under the hallucination that her house was on fire.

In these instances, the judgment was clearly misled, warped, or in abeyance, and the conduct governed by a belief in the hallucination. I could multiply cases, but to do so is unnecesary, as the origin of delusion in either illusion or hallucination is sufficiently clear.

Delusion must, however, be carefully studied and analysed, because it is the most conclusive evidence of insanity, and the evidence most favourably received by lawyers, and in courts of justice, and by those in authority. But delusion is not always indicated by verbal expression; it is often shown in peculiarities of manner, and incongruities in actions or in dress. These peculiarities, when analysed, sometimes, appear to spring immediately from an illusion or an hallucination, at other times they are intimately associated with disturbance of an emotion, very commonly the emotion of self. Thus, self-conceit or self-depreciation are frequently associated with the delusion, or the spring of the peculiarity in conduct.

A delusion is often first indicated by an unusual restlessness in manner, which is the outcome of the disturbed emotion, whilst the meaningless inquisitiveness or listlessness, which frequently accompanies this restlessness, indicates that the faculty of attention has, in some degree, been interfered with. A fixed vacant stare often indicates that the patient's attention has strayed beyond the normal attraction of ordinary vision, and an erect unyielding gait, or an assumed air of importance is the common expression of delusionary exaltation. Measured strides in walking, or any excentricities when they appear as changes in conduct, deserve to be noted. I knew a patient who exhibited his departure from healthy thought, by picking out and walking upon particular patterns on the carpet. But all departures from ordinary conduct should be studied. Any ridiculous conduct, which the person fails to recognize as incongruous, may indicate insanity, and any undue assertion of self, in however small a degree, may indicate an unhealthy centring of thought in self, and as self-consciousness is the last faculty that can be lost, it is the first to stand out in prominent relief when the others begin to fail; it is then but a step for self to become the prevailing idea, for the attention then becomes directed to self to the exclusion of everything else. The patient becomes imbued with the impression of his importance, his wealth, his ability, his power, or his wickedness and his shortcomings, his poverty, and his misery, and, in consequence of the lesion of attention, always more or less present in insanity, he cannot retain a correcting impression sufficiently long to learn from it that his self-conceit or his self-depreciation are untrue, and, in consequence, he is not able to see the absurdity of the outcome of his delusion when it renders him ridiculous.

I have known a patient to sit upon every chair in the room in the space of five minutes, examine every article within reach, stare at an object and be unable to say what he was looking at, and walk with an importance worthy of a beadle of Bumbledom, and all this without any expression of a delusion. A delusion of greatness was, however, underlying all, which, if put into words, might be described as an over-weening confidence in self, or a disturbance of the emotion of self, coupled with delusion of greatness.

The impulse which permits violence arises from one and the same spring. The common lesion of attention abolishes the power to check or control, and the promptings of self-conceit or self-depreciation are acted upon instantaneously.

The peculiarities of manner which spring from an overbearing self-conceit are the most constant and striking. The lunatic who has a delusion of his own mighty power, will take means to show that he is powerful, and it is out of this common delusion that much of the violence displayed by the insane arises. Incongruous hilarity or unwarrantable dejection, shown in unseemly dancing, singing, or laughing, or uncalled-for weeping, or despondency, is sometimes a prominent expression of a delusion, which, when analysed, will usually be found to spring from self-conceit or self-depreciation.

Suspicion or distrust of all but self and disaffection towards relatives and friends is but a natural sequence of the undue prominence assumed by the emotion of self. Sometimes distrust and suspicion are the most evident manifestations of departure from a healthy state of mind, and they prompt very incongruous acts. I knew a clergyman who left the pulpit, on one occasion, in the middle of his sermon from a suspicion of his wife's infidelity. A moment's reflection in a sane mind would have belied the suspicion in this instance. The patient had never shown any indication of insanity before, he is now in an asylum and when I last saw him he assured me that he was "God Almighty."

Peculiarities in expression, are definite enough in the declaration of absurdities, in the loss of personal identity, and in the expressed belief in incongruities which divert the conduct from a rational course: thus, a lunatic will have the delusion that he is "God" or "Christ" or "a King" or "the Devil," and if he give verbal expression to his belief, the delusion is definite. The effect of delusion upon the conduct of course varies, and it varies with duration; sometimes, instead of prompting to the assumption of an air of importance or violence, the patient will be so impressed with his belief that he will dismiss all other considerations. He will be content with declaring himself "God" or "Almighty" and will eat his dinner expressing regret that his present circumstances will not permit him to display his power. A patient in St. Luke's Hospital used to style herself the "Queen of the weather." She was usually satisfied with the declaration that she sent wet weather as a judgment, and she used to ask for thanks whenever it was a fine day. But slight peculiarities of expression, in particular and individual cases, as indicated in a change of language or a departure from an ordinary vocabulary, may be evidence of insanity, and spring from a delusion, otherwise undefined. A gentleman I knew of, whilst lecturing one day, suddenly ceased to speak in English, and continued his discourse in French. It then became evident to his hearers that his mind was off its balance. He continued insane for a long time, but I believe afterwards recovered. If a man of devout and unblemished character, "an eminent Scotch divine" for instance, suddenly commences to use foul language, you may search for a delusion, and will probably find it; this is the kind of case in which swearing may be an evidence of insanity. If, as a similar instance, a lady of refinement and education suddenly commences to swear or use foul language, the fact will indicate that her mind has left its ordinary channel, or at all events, that it is no longer under its ordinary check, and this loss of control should be your search warrant for evidences of delusion.

Peculiarities of dress, exampled in such absurdities, as imitation of regal robes, and dresses of state dignitaries, are obvious enough. But the subject of delusion will dress in a variety of incongruous ways: a man will turn his coat inside out, or button it down his back, or a woman will cover her bonnet with various pieces of ribbon and put it on upside down, or deck her hair ridiculously, or she will pin an apron behind her, or commit some other extravagance in dress, which, appearing as an incongruous change from ordinary deportment, must be analysed for a delusion. Dr. Blandford mentions the case of a clergyman who was recognised to be insane by his wearing a white hat; and you might as readily arrive at a conclusion as to insanity, in a particular case, if you saw a man without a hat at all.

One of the strongholds of popular error, on the subject of insanity and delusion, is, that the lunatic is unable to give utterance to anything but nonsense, or that he is insensate. The patient labouring under the delirium of fever is more or less insensate, and the lunatic, is often unreasonable in consequence of the delusions, and, when excited, he is often so overwhelmed by the delusion, that all other ideas are excluded from his mind; but in calmer moments he is not insensible of that which goes on around him, the maniac and melancholic can speak of their families and their friends, and upon most other subjects, without influence from their delusion. The passively demented become more or less insensate, but they often, on recovery, are able to tell you much that has passed during their alienation, so that even they are not altogether unimpressionable.

Delusion, though the commonest expression of

insanity, is by no means a constant or a necessary accompaniment of the insane state, and after all it is but one of many symptoms of cerebral imperfection. It may, however, be regarded in the light of a standard wherewith to measure the degree of perversion of brain function.

In many cases of melancholia there is no apparent delusion; likewise in dementia, delusion is often markedly absent. Again, we have seen a disturbance of mind in fever which is often unaccompanied by delusion; often, however, in fever, there is illusion or hallucination of a passive kind, which is termed delirium, it might be called acute delirium.

In delirium tremens you have a busy delirium; the patient labours under illusion or hallucination of a very imperfect kind, and though his phantasmagoria may produce a feeble delusion, so as to somewhat govern the conduct, it is usually of a more or less gentle character, and the patient, as a rule, is very easily persuaded.

Delirium, as defined from delusion, is a constant wandering of the mind, without any impression becoming either corrected or fixed. An act committed in an acute delirium is committed without a definitely fixed motive, whilst that committed under a delusion, which has been regarded as a chronic delirium, is usually performed in consequence of a very definite motive, though that motive is the outcome of disease and perverted function.

CLASSIFICATION.

It is necessary to examine the various classifications of insanity that have been proposed, though I am strongly of opinion that we can hardly desire a better than the ordinary and received natural classification which I have given you; at all events, with slight modification our natural classification is essential, and it is in general use. It is true, that it only embraces intellectual insanity, and that it is essentially a classification of symptoms, but insanity *per se* is only symptomatic. Regarding a healthy mind as a physiological phenomenon of healthy brain, insanity is the pathological phenomenon of the morbid brain, or a perversion or imperfection in the ordinary physiological phenomenon, in consequence of imperfection or unhealthiness in the brain tissue.

Clinically, we have to regard the case from its symptomatic aspect, and we can add to the classification of symptoms, from intellectual expression, those which are the outcome of the emotions. There are, it is true, certain states, such as some of those included in the class which has been termed moral insanity, which cannot easily be recognised under such heads as mania or melancholia, but they may come under the head of either dementia or amentia; moral insanity being frequently moral imbecility, whilst amentia and dementia are the two great sub-classes into which insanity is naturally divided. The various pathological conditions of brain and cord may be classified together under heads, as atrophy, sclerosis, granular degeneration, pigmentary degeneration, and the like; but such pathological grouping does not belong to mental disease. Any pathological change in the brain may produce dementia, or, if congenital, amentia, there being no special pathology of mania or melancholia. When, however, it is clearly recognised that all insanity is the phenomenon of perverted function, and is a consequence of morbid states of tissue, the grouping together of appearances of the phenomena will not be difficult.

The classification which was proposed by Esquirol I have had drawn out in the table on the wall.

Esquirol's Classification.

I. Conditions of Depression.
    1. Hypochondria.
    2. Melancholia.

II. Conditions of Exaltation.
    1. Acute Mania.
    2. Monomania.

III. Conditions of Mental Weakness.
    1. Craziness or Incoherence.
    2. Dementia and Fatuity.
    3. Idiocy and Cretinism.

IV. Paralytic Dementia.
    General Paralysis of the Insane.

It is obviously incomplete and is rather an enumeration of certain insane conditions, than a classification.

A classification proposed by Dr. Maudsley is more complete though its basis is metaphysical, and it is not always easily applicable in daily practice. It is contained in the tables on the wall.

MAUDSLEY'S CLASSIFICATION.

I. *Affective or Pathetic Insanity.*
  1. Maniacal Perversion of the affective life. *Mania sine Delerio.*
  2. Melancholic depression without delusion. Simple melancholia.
  3. Moral alienation proper. Approaching this but not reaching to the degree of positive insanity is the insane temperament.

II. *Ideational Insanity.*
  1. General.
    a. Mania. ⎫ Acute
    b. Melancho- ⎬ and
       lia. ⎭ Chronic.
  2. Partial.
    a. Monomania.
    b. Melancholia.
  3. Dementia, primary and secondary.
  4. General Paralysis.
  5. Idiocy including Imbecility.

A classification which has attracted considerable attention, but one which has been severely criticised, is, that of the late Dr. Skae. It is based on etiology, and was first published by him in his address to the Medico-Psychological Association in 1863. It has been slightly altered, or rather added to since its first publication, and now is arranged as follows.*

Insanity with Epilepsy.
    ,,   ,, Pubescence.
    ,,   ,, Masturbation.
Satyriasis.
Nymphomania.
Hysterical Insanity.

Amenorrheal Insanity.
Post-Connubial Insanity.
Puerperal Insanity.
Insanity of Lactation.
  ,, ,, Pregnancy.
Climative Insanity.

* Morrisonian Lectures for 1873. Delivered by Dr. Clouston. See *Journal of Mental Science*, Oct. 1873.

Ovarian Insanity.
Hypochondriacal Insanity.
Senile Insanity.
Phthisical Insanity.
Metastatic Insanity.
Traumatic Insanity.
Rheumatic Insanity.
Podagrous Insanity.
Syphilitic Insanity.
Delirium Tremens.
Dipsomania.
Insanity of Alcoholism.
Malarious Insanity.
Pellagrous Insanity.
Post-febrile Insanity.
Insanity of Oxaluria or Phosphaturia.
Anæmic Insanity.
Choreic Insanity.
General Paralysis with Insanity.
Insanity from Brain Disease.
Hereditary Insanity of adolescens.
Idiopathic Insanity {Sthenic. Asthenic.

But this classification though highly valuable as an enumeration of causes, is manifestly imperfect, and I think can never claim to be more than an enumeration; it cannot be a classification, it is wanting in scientific exactness, since it gathers together under the head of idiopathic insanity all forms that cannot be referred to one of the causes mentioned, and this idiopathic insanity necessarily includes the greatest number of cases. But a stronger ground of objection to its practical application is, that any of the forms of insanity, enumerated in Dr. Skae's classification, will assume one of the appearances recognised under the natural classification, though sometimes the same case alternates between one expression and another. Thus, a case of insanity dependent upon gout, may at one time appear as mania, at another as melancholia, at another as dementia, but the expression of the insanity is not modified by cause, unless the modification be one of degree (but this is usually slight). Mania, melancholia, &c. are definite expressions of insanity, and require to be considered as the symptoms of insanity—" a perversion of the cerebral function of mind from temporary or permanent change in the brain tissue"—apart from the special cause in each case. Undoubtedly the special cause should be sought for in every case, for it is highly important as a guide to special treatment, but the classification of insanity must be arranged from a natural relationship of its symptoms; but a slight modification from the grouping which I have given you is, as I said, necessary. The expression " General paralysis of the insane" must be omitted from the list, for the insanity associated with progressive paralysis is always either mania, melancholia, or dementia, all of which may be considered under the sub-class dementia. Idiocy and imbecility should be considered under the sub-class amentia. It will be found that all cases of insanity not amental, are in a greater or less degree, either maniacal, melancholic or passively demental, or perhaps alternately any two or all three, and any, or all of these expressions may be present with or without definite delusions, and as mania and melancholia are in a greater or less degree dementia, we might state the classification thus:

Insanity
- Dementia
  - Mania. *Excitation.*
  - Melancholia. *Depression.*
  - Dementia. *Passive State.*
- Amentia
  - Idiocy.
  - Imbecility.

But insanity so classified, refers only in its general application to disorders immediately involving the intellect, whereas insanity does not always involve the intellect. The emotions may be in a condition of insanity, either in conjunction with the intellect or alone, and a perfect classification must embrace them also. Thus we have—

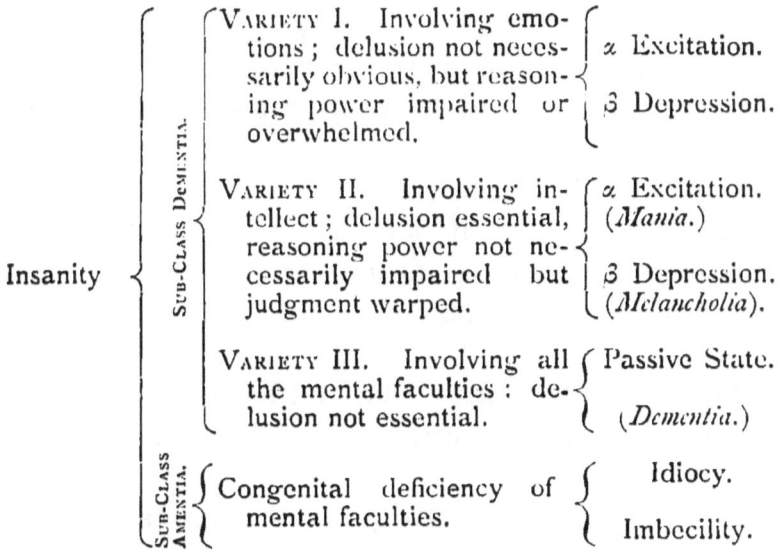

Insanity
- Sub-Class Dementia.
  - Variety I. Involving emotions; delusion not necessarily obvious, but reasoning power impaired or overwhelmed.
    - α Excitation.
    - β Depression.
  - Variety II. Involving intellect; delusion essential, reasoning power not necessarily impaired but judgment warped.
    - α Excitation. (*Mania.*)
    - β Depression. (*Melancholia*).
  - Variety III. Involving all the mental faculties: delusion not essential.
    - Passive State. (*Dementia.*)
- Sub-Class Amentia.
  - Congenital deficiency of mental faculties.
    - Idiocy.
    - Imbecility.

The intellectual and emotional disorders are commonly blended together, though sometimes they are quite distinct. It is not uncommon to see emotion in abeyance by reason of insanity involving the intellect, as the loss of affection towards children or towards parents, but we also, sometimes, see in what is termed moral insanity the involvement of the emotions alone, the intellect being more or less undisturbed. In the abstract, illusion, hallucination, or delusion, are present in insanity involving the emotions, since the emotions rank among the feelings, and, as a sense, are subject to mistake.

The mind ultimately is resolved into senses, emotions, and intellect:—volition being an acquisition resulting from various influences,—and insanity starting from the senses, displays itself in perversion of either the intellectual or the emotional part of mind, or of both. The acquired faculty of volition is under all circumstances obedient to the promptings of the intellect, or of the emotions, out of which it takes its origin, but insanity does not take its origin in volition.

Insanity of will, if such there be, is therefore clearly a secondary condition, and dependent upon the promptings of perverted intellect or perverted emotion.

We may enumerate in tabular form the varieties of insanity as they spring from perversion of the senses.

Insanity, (taking its origin in the senses) when it appears as

Illusion—is usually optical or auditory, but it may be of any sense. And when appears as

Hallucination—though it is most commonly associated with sight and sound, may be, and often is, of taste, and smell, and of the tactile sensation generally.

We may get a comprehensive view of the basis of insanity by regarding the mistakes of sense, and the association of illusion or hallucination with delusion, in a tabular form. The arrangement though necessarily imperfect is in part modified

CLASSIFICATION.

from Professor Laycock's classification of morbid instincts.

Illusions or Hallucinations of		Delusions		
		Perverting the judgment or originating mistakes,	and	governing the conduct.
Sound	give impressions of	Voices and Mandates,	and often incite	Homicide, Suicide.
Sight	,,	Visions,	,,	Pamphobia, Homicide or Suicide.
Smell	,,	Fanciful Odours,	,,	Violence. Pyromania.
Taste	,,	Morbid Taste,	,,	Belief in the pleasantness of nauseous substance, or the reverse, as an impression that food has become poison; readiness to take nauseous substances in the mouth; refusal of food; violence.
Feeling without emotion	,,	Morbid Muscular, and Cutaneous Sensations,	,,	Abnormal muscular sensations described as creeping, and the various conditions described as auræ, as observed in epilepsy. Desire for stripping or for excessive clothing.
Feeling involving the emotions	,,	1. Morbid Alimentary Sensations,	,,	Fear, and belief in the presence of various foreign bodies, or of living creatures in the abdomen; refusal of food, or excessive feeding to nourish contained animals.

Illusions or Hallucinations of		Delusions		
		Perverting the judgment or originating mistakes,	and	governing the conduct.
Feeling involving the emotions	give impressions of	2. Alimentary necessity,	and often incite	Pica, Cannibalism, Gluttony, Bulima, Onomania.
		3. Sexual necessity,	,,	Erotomania, Nymphomania, Satyriasis, Onanism.
		4. Domestic disaffection,	,,	Groundless Hatred, Misanthropy, Infanticide, Paricide, Matricide, &c.
		5. Personal misconceptions,	,,	Pride and Exaltation, Self-mutilation, Pamphobia.
		6. Social disregard,	,,	Kleptomania, Chicanery, Accumulating, Hoarding, Love of low society, Vagrancy, and tendency to Vagabondise.

Insanity involving the Emotions.

Insanity involving the emotions primarily, may give rise to any of the forms of disease mentioned in the table, but it is often exhibited in uncontrolled vagaries of will, accompanied with many of the well-known symptoms expressed under the terms hysteria, and hypochondria.

There are in fact a large class of disorders, in which the emotions are disturbed, which border very closely upon, and sometimes run into, insanity, and which I believe are associated with abnormal conditions of the sympathetic.

In briefly considering this subject I should premise that depression implies a more or less definite pathological condition; at all events in the sense that it implies a lessening of vitality, all pathological states being coincident with a lesser degree of life. Excitement is no less pathological; it is the rapid and uncontrolled performance of a function, the tendency of which is the rapid exhaustion of the organ upon which the function depends.

You should remember that all functional disorders are dependent upon pathological states of organs. The expression *functional disturbance* has long been used too vaguely; disturbance of function is as strictly pathological as normal function is physiological, though the disturbing cause may be slight, or temporary, or easily removable. The affection called Hysteria is one of those which, often, is designated "functional disturbance," but it is no less a neurosis than insanity, which might equally be termed functional disturbance.

That emotional disturbance is dependent in a great measure upon depressed states of the sympathetic is, I think, clearly shown in an analysis of the phenomena associated with disturbed emotions. In the so-called "hysteria," for instance, we notice undue expression of emotion, undue laughing, or undue crying, which the patient cannot control, and with them we frequently see disturbance of the digestive organs, loss of appetite and flatulence, irregular beating of the heart, and an unusually large excretion of urine, but all these are the immediate result of a depression of the sympathetic, whereby its controlling influence has become enfeebled.

But without seeking for evidence in abnormal states, we find in the physiological phenomenon of blushing a strong confirmation. Blushing is a purely involuntary phenomenon, a purely emotional expression, of a painful character, and dependent entirely upon the sympathetic; the cause of the depression being the more evident when we call to mind the psychological principle* that "states of pleasure are concomitant with an increase, and states of pain with an abatement, of some, or all, of the vital functions." But in this we find an intimate association though withal an independence of function between the brain and the sympathetic. We may see the same in shock, and we may learn from shock many a valuable physiological lesson. Shock produces precisely the same effect, whether the result of a physical injury, or of an intellectual blow, a moral shock, as it is termed, as exemplified in the communication of unexpected and painful intelligence. Both causes of shock at once depress the sympathetic; sometimes a sharp epigastric pain is felt; in both, the condition of vitality is lessened, the circulation primarily is disturbed, the heart may beat very slowly or very quickly, but usually it beats labouredly and irregularly, and various emotional expressions will be manifest. In extreme physical

* *Mental and Moral Science*, by Alexander Bain, M.A., p. 75.

pain the lacrymal glands will become hyperæmic, and tears will flow; this is from loss of control through depression of the sympathetic. In the shock from causes operating through the mind, i.e. through the intellect or through impressions upon the surface of the brain (moral causes), the depression is often greater, the base of the brain becomes congested as well as the lacrymal gland, and sobbing is a consult result. But the effect is further shown by depression of other sympathetic functions; thus the appetite is lost, and the person feels sick or perhaps faint. The faint sensation is a very definite evidence of sympathetic disturbance, for it results from a lessening of the amount of blood circulating in the brain, through an irritable condition of the cerebral vessels, the nerves of which, as you know, are derived from the sympathetic. In a greater degree this would produce actual unconsciousness, as appears in the cases of epilepsy from frights. These are clearly emotional causes depressing the sympathetic, the direct effect of which we see upon the cerebral circulation through the vasi motor nerves.

A familiar illustration of the effect upon the brain's circulation, is afforded by the not uncommon experience of a student who witnesses an operation for the first time; he feels sick, and faint; but many other people, who are unsteeled to the sight, become actually unconscious at the flow of blood.

The most severe cases are those which I mentioned under the head of melancholia, in which the patient never recovers from the emotional shock, and either dies outright or else becomes permanently insane. But in a minor degree the influence of depressed states of the sympathetic, from emotional causes, may be very perfectly observed in the asomnia which excessive grief often induces.

In the normal condition, sleep is the result of a reduction of the circulation in the brain, through a lessening of the calibre of the cerebral vessels, the effect being brought about by the cervical ganglia of the sympathetic. But when excessive grief overwhelms a person the sympathetic may become paralysed, and asomnia results, which, unless speedily combatted, will be followed by insanity, (disturbance of cerebral function,) or by coma and death. The same effect is sometimes produced, though usually in a minor degree, by continued irritations, vexations, annoyances, or by overtaxing the mental powers, with subjects wherein personal feeling more or less operates. The effect of worry, anxiety, and overwork of the brain, is to upset all the functions which are under the control of the sympathetic; digestion becomes impaired, the appetite is lost, the tongue becomes loaded, the bowels irregular, the bile either deficient or highly pigmented; the pulse usually becomes hard and irregular, and the heart liable to palpitation; indefinite pains fly about the head, chest and limbs, the urine is either considerable in quantity or highly acid, and sleep becomes imperfect, or broken with phantasmagoria and horrid dreams. The patients complain till the story of their ailments appears as an idle tale: no disease of any viscera is apparent, nor do they often waste in any great degree, though the tissues sometimes become flabby; nothing is made out of physical examination unless it be an anæmic bruit; the patient becomes the prey of undue fear, and his condition is miserable in the extreme. The misery, indeed, that these patients suffer, is almost beyond description, and sometimes the cardiac disturbance is so great, that the patient will convince himself that he has heart disease, and he will go about in constant apprehension lest he should die suddenly, and in truth he will often believe himself to be dying. He will alarm the whole household in the night, from a feeling that dissolution is impending, but on the morrow he, or she, is almost well, and at length you are inclined to regard the cry as that of "wolf," "wolf." But they suffer agonies both mental and bodily, and though in the presence of other people they usually seem better or feel better; their condition is, nevertheless, one which needs and will be improved by treatment. The state is sometimes attendant upon, if not actually brought about by physical disease, particularly morbus cordis. A soldier who has fearlessly braved the cannon's mouth, may become the subject of morbus cordis, and suffer from such an amount of depression, and emotional disturbance, that groundless fear will completely shake his confidence in himself and he will become the most miserable of beings. But in a large majority of such cases, as in all cases of so-called hysteria and hypochondria, there is probably an original defect in the sympathetic, which is made manifest by emotional disturbance, whenever the patient labours under a sufficiently depressing cause. Very slight physical suffering is often sufficient to develope distressing emotional symptoms, and these are always associated with disturbance of those organs, which are more directly under the control of the sympathetic than of the brain. These general facts at least seem to fix the seat of the emotions in a great degree upon the sympathetic, although changes in the sympathetic have not yet been sufficiently observed. The pigmentation of the sympathetic ganglia, which I have already mentioned is, however, a step in the right direction, and I have no doubt further observation will furnish us with some more startling facts.

The popular notion that the will is primarily at fault in the neurosis called 'hysteria,' is incorrect. The patient's will is overwhelmed by the excess of emotion, and the notion that such patients have "only to make an effort" is as ridiculous as the name, which refers the disorder to the uterus. The will, of patients suffering from sympathetic depression, is held in abeyance by the emotional disturbance; the temporary amaurosis, ptosis, hemiplegia, paraplegia, aphonia, dysphagia, and all the other anomalous symptoms which these patients suffer from, are all states which he or she would fain be rid of, but the effort to shake them off, is as impossible as would be an attempt to walk by a patient suffering from acute rheumatism or gout. To urge a patient suffering from depression of the sympathetic to make an effort to overcome the emotional disturbance, is I think wrong in practice, and I believe much more may be done with rest; and, sometimes, benefit may be derived from such drugs as iron, bromide of potassium, and vegetable neurotics, and in particular I should mention digitalis in small doses. In the early conditions, sleep should be induced by any means within reach, as with warm baths, chloral, and alcohol, and the patient should, as far as possible, be released from all causes of anxiety and worry, and from mental and physical labour.

Moral Sense and Moral Insanity.

There is no abstract idea of right and wrong unless it be that right is that which is the greatest good to the greatest number, but the opinion as to what is the greatest good to the greatest number varies in different countries, and therefore the code of morality in different countries varies, and this variation is so great that what is considered in one country as crime, in another may be regarded as honourable. As a matter of fact, the morality of a country is defined by the opinion of society in that country, and the code becomes something to be learnt by individuals. What we call conscience, and which indeed is the only test of a moral sense in an individual, is but the standard of right and wrong formed by experience in the mind of that individual. Almost the earliest impressions attempted to be instilled into the infant mind, by its fondling mother, are arbitrary separations of right and wrong as defined by her own conscience, and she endeavours to separate in the infant mind, on two sides of this hypothetical standard line, ideas of right and wrong which are often most puerile, and sometimes not antithetical. As the child grows, and the basis of his experience enlarges, his standard of comparison advances: many of the puerile wrongs of infancy and childhood appear as wrongs no longer, and his line of separation, *i.e.* his conscience, becomes fixed in accordance with the moral and civil laws of the polity in which he is placed.

The question is often asked in this form, Have we an innate sense of right and wrong? and I think insanity answers the question conclusively in the negative. I have stated that some maniacs retain the power of reasoning in a remarkable degree; and numbers of lunatics, particularly maniacs, are singularly conscious of the difference between right and wrong; the answer to the question, " Does the patient know the difference between right and wrong?" depends, therefore, in some measure, upon whether or not he be the subject of moral insanity or moral imbecility. A patient may suffer from mania or from melancholia which may assume the form of violent frenzy, during which he will perfectly discriminate between right and wrong, whilst the morally insane has lost the discrimination, or the morally imbecile, will never know the difference. The moral imbecile is incapable of learning morality, *i.e.* his brain is unimpressionable, and you cannot develop in him a moral sense. In illustration, take an example from among the idiots, you can teach many of them to work, but you cannot teach them that it is wrong to lie and cheat.

The development and decadence of the moral sense you may best see in what may be termed the higher stages of cultivation, and as illustrated by chastity.

The codes of morality which have placed chastity in the highest position is the code of the highest civilization; and in this country, where chastity takes a high position in the ranks of honour, we notice a very great disregard of it among the uneducated classes. If we then regard the insane, who are by degeneracy two or three grades removed from the uneducated, we shall constantly find the gravest departures from societies, or, as they are called, the moral laws of chastity.

If, for instance, we find a woman of delicate cultivation and moral training suddenly become indecent in act and speech, we must conclude at once that some change has occurred in her moral sense, and we can only regard it as a change in her brain, whereby the reasoning power has been disturbed, and we may at once look for other evidences of insanity. When a change, tending to retrogression, occurs in the brain, the sexual appetite asserts itself, the power of control is gone, the sexual appetite is strong, and the woman asks herself why she should not gratify it. She fails to see society's reason, and with this loss of recognition of wrong she has lost what is called the sense of shame and remorse. She indulges her sexual appetite, regardless of consequences, and with satisfaction, boldness and temerity, rather than with the trepidation and shyness, which would be present with a woman who errs from confidence, perhaps undue confidence, in the man to whom she surrenders her charms. If you now take the case where the brain was never capable of receiving the impression of the moral sense at all, and keeping the instance of chastity still as our example, you will find cases, both among children and adults, who are incapable of realizing society's restrictions on the score of chastity, and who yield themselves up to sexual impulse because the appetite calls for satisfaction; there is no restraining influence or impression in their minds to check the impulse, nor are the brains of these individuals capable of receiving such an impression. Dr. Maudsley has quoted several most interesting and instructive cases in his *Physiology and Pathology of Mind*, occurring in both adults and children, and of the latter he remarks, " The afflicted child has no true consciousness of the import of its precocious acts: certain attitudes and movements are the natural gesture-language of certain internal states, and it is little more than an organic machine, automatically impelled by disordered nerve centres."

If we recognize the development and the degeneracy of the moral sense in the case of chastity, we have but to look a step or two below to see the retrogression in other departures from moral rectitude. I need not again refer to the case of drinking, and the loss of all shame and control in regard to it; but the case of lying, cheating, and pilfering, as sometimes seen in children, and again more strikingly seen in adults who have led unimpeachable lives, and then become insane, furnishes the most conclusive evidence of the ordinary cultivation of the moral sense, and its retrogression with degeneration of brain tissue.

In the child wanting in moral sense, it is usual to find hereditary nervous defect. You cannot impress such a child with the idea that it is wrong to lie, or cheat, or steal. He will take what he sees, and no questions of right or rights will influence him more than, or perhaps as much as, they would influence a hungry dog, when within reach of a lump of pudding. Such a child might tell a lie about the theft, not because he is conscious of having done wrong, but because he has learned from experience, that if he tells the truth he will get a thrashing, (a dog, save for speaking, will learn as much,) or the child will tell the lie, because experience has afforded him pleasure in seeing the perplexity of those to whom he tells it. If detected he will often candidly avow the theft shamelessly, declare that he sees no wrong in it, and though you thrash him within an inch of his life he will only feel the stick, but not remorse.

One step further, and you have complete moral imbecility in which the miserable patient is akin to the wild beast, or certainly a little lower than a domesticated animal. Such as these will commit murder without realising the act to be wrong, or in the least degree comprehending the enormity of the crime. Nor can they ever gain the impression that there can be a wrong in anything that which their impulse prompts them to do. It is not criminality ; the true criminal feels conscious of the wrong, and feels the punishment, and often remorse and shame. The moral imbecile is mentally unimpressed by the punishment, is unconscious of the wrong, feels no remorse nor shame, and, probably, has dismissed into oblivion and for ever, all thought of the act for which he is punished. Often, indeed, he has almost forgotten the act long before the punishment, meeted out by reason of the existing state of our law, begins.

Much has been written, and more has been said upon the subject of moral insanity. The difficulty is most felt in courts of justice; but whilst two great professions, Law and Medicine, are at issue on the most elementary portion of the subject of mental disease, it seems hardly possible that they will speedily agree on that part which perhaps of all others is abstract.

Law, which professes to be the perfection of reason and common sense, admits, as I mentioned in our first lecture, insanity as annihilating criminal responsibility, only when it can be shown that the subject of such insanity does not know right from wrong; and it is, indeed, a curious fact, but the same law which adheres closely to the metaphysical boundary in the definition of insanity in criminal cases, will admit almost any vagary in the determining of lunacy in a civil case, as though the sordid coin of a miserable lunatic had more claim upon the protection of the state than the life and honour of a human being, and the sacredness of the feelings, reputation, and interest of his family as citizens of the world.

But a court of justice offers the worst arena for arguing such a question in. The position of counsel as compared with that of witness, is, by virtue of his office, one of advantage, whilst from the subtleties of metaphysical argument, the questions cannot always be entered upon, even by the experts, in a law court, with any sort of hope, that agreement will result from the discussion.

The case of Christiana Edmunds furnished a striking instance of the difficulty experienced by lawyers and doctors in their attempts to reconcile philosophy and law. The reason, perhaps, being that the former advances by greater strides than the latter; philosophy being ever ready to extend its landmarks, whilst precedent seems to be the watchword of the law.

In the case of Christiana Edmunds, there was overwhelming evidence of hereditary predisposition, and the utter disregard of the consequences of spreading death broadcast, was, in itself, presumptive evidence of mental disease; whilst the evidence which was adduced at the trial, clearly exhibited a mind actuated and moved by one idea only, and an idea which ruled so powerfully that all correction or reasoning as to consequences was impossible, even had she at any time known how to distinguish right from wrong. After her trial the same ruling idea held her mind. She had a morbid unreasonable and unreasoning love, she loved something or somebody, and every obstacle between her and the object of her love, must be cleared away at any cost. Thus a morbid impulse urged the commission of unreasonable acts without the least regard to consequences.

This is the common history of moral insanity, when associated with erotomania.

The unfortunate subject of moral insanity, differs from the ordinary criminal, who deliberately chooses the wrong and risks detection, in that the morally insane person has no choice but to do the wrong, being impelled by a dominant and ruling idea, which no amount of reasoning can set aside. The individual conceives that something must be done, and in order to allow him to accomplish that something, every other consideration must give place. The act often bears a strong resemblance to an act of reason and deliberate criminality, but an analysis of it will generally show it to be akin to the act of delusion. The brain which gains the impression, "I, A love B, and must destroy everybody and everything between B and me," and is unable to reflect and reason upon the consequences of the act or acts of destruction, is in much the same condition as the one which assumes the impression, "I am a king and must get a crown;" in the first case the surroundings are as incapable of correcting the impression as in the second, and yet in the second, the public unhesitatingly declare the subject to be mad; whilst in the first, because something ludicrous is not apparent he is declared criminal. The same incapacity for reason constantly occurs in what is termed kleptomania. The subject's brain in kleptomania has one prominent idea, "I must take," or in pyromania the idea "I must set on fire" is dominant. No amount of teaching or arguing would overrule these ideas; the unfortunate patients steal by reason of the impulse, and without object take what is no use to them, or set a house, or a large or small article on fire, from no spite, or for no reason, but that the idea of taking or setting on fire is dominant, and in their morbid minds there is no power of correction. In consequence of the insane act many insane persons are convicted of theft or arson and are incarcerated in jail, when they should have received all the kindness, tenderness, and care that medical treatment and skill could give them.

Moral insanity is very often hereditary, and the unhappy inheritance has a claim upon the consideration and the sympathies of all. As the branch resembles the tree so the child resembles the parent, and if the father or mother has lacked moral sense, it is not surprising that their children should be deficient in the same. If a blind man has a blind son, no one is surprised, and if the moral sense were clearly recognized by lawyers and the law, its annihilation would be no less easily recognized than annihilation of the sense of sight, and still less would its absence be open to question when hereditary cause was shown. A man may be born with a defect in vision, and in like manner a man may have defect in what is termed mental vision. All he sees and appreciates mentally is askew, and all he does is askant, and it would be cruel to hold him responsible for that over which he has no control. If I grant the proposition of Mill that all our knowledge is made up of our experience, I will claim that which Mill did not state, namely, that our knowledge will be modified by the capability of our brains for receiving impressions. A person with obliquity of vision cannot receive visual impressions in the same manner as a person with a perfectly straight vision; and by parallel condition, in the case of moral sense, certain brains are incapable of receiving truthful impressions, and so forming the ordinary standard opinions of right and wrong. We should consider also the case of the many maniacs, in whom the moral sense is often markedly absent. The victims of mania are sometimes notable for their chicanery, and their habits of hoarding, and pilfering; it can, therefore, hardly be a matter of surprise, if the same propensities appear, at some time, in the child of such a man.

Unfortunately for the victim of moral insanity, the malady is not recognised or observed, until some overt act is committed, and then to the public, who can only judge from the naked facts which are placed before them, declare the miserable creature to be a criminal, and the absence of a history of previous insanity tends to negative all argument of insanity that may be urged in the prisoner's favour. If one inmate of a lunatic asylum murders another inmate of the asylum or murders his keeper, public opinion at once acquits the murderer; but if a lady steals a silk handkerchief out of a shop, or a schoolboy possesses himself of another boy's books, or toys, or clothes, or money, society will scarcely, if at all, admit the plea of kleptomania, however clearly it may be demonstrated, and however feasible, in the case in which it is advanced. Still less will society be willing to excuse, when the strength of the evidence of moral obliquity rests upon insanity in a former generation.

A lady called upon me one day, not long since, and asked my advice regarding one of her sons, on whose behalf she had many times consulted me, and upon a suggestion I made she burst into tears, saying, "I am sorry to tell you that my son is a thief." Now this boy had been most judiciously and carefully brought up. His parents were devoted and loving, and he suffered neither want, cares, nor anxieties; had received the education of a gentleman, was provided with all necessaries, not stinted in pocket money, and yet he could not be trusted. I learned from his mother, that he would steal money whenever he had the chance, and squander it away in wanton extravagance.

The history is soon told. His father had been a patient in an asylum, the subject of melancholia with delusions; a sister had shown maniacal symptoms, and this boy, of fair intellectual capacity, was of so warped a moral sense, that he could not appreciate the distinction between mine and thine, nor could the appreciation have been beaten into him by means of the stick.

Another case about which I was consulted, was

the son of a gentleman in a high position, and the boy was sent down from one of the public schools because he stole. Among the things he enumerated to me, were blankets and sheets which he took off his own and other boys beds, clothing, and a variety of articles amongst which were two pairs of plated nut-crakers which he sold for a shilling a pair. On asking him if he wanted the money, he said no, that he had plenty of money, but that he felt something he could not describe, urging him to take the things. He was greatly distressed, but he declared that he could not control himself, so irresistible was the impulse, and he then admitted that the day before I saw him, he had stolen some jewelry, the property of his father, although when first charged with the theft he stoutly denied it. His impulse was simply to take; after he had possessed himself of many of the things, he did not know what to do with them, and attached no value to them, the jewelry he stole he simply treated as toys, some of the articles he stole at school he hid in a garden, and others he sold; it appeared that he disposed of them merely to rid himself of their encumbrance, he felt that the things had been stolen and must be got rid of in some way, and the readiest means was that of selling. In this case I failed to find any previous history of insanity, or to discover any satisfactory cause for his imperfect moral sense. But we must remember that it is often impossible to get a truthful history from relatives.

A case was related to me by a clergyman who had under his charge a boy who developed a propensity for stealing money, but who nevertheless stole so cunningly, that his tutor was unable to bring any charge home to him. It was, however, discovered that the boy used to talk in his sleep, and the clergyman therefore addressed himself to the youth during the hours of sleep, and asked him from whom he had stolen the money, the various sums he had stolen, and what he had done with the cash. The boy answered every question, stated the sums accurately, the manner in which he taken them, and named the place in which he had secreted the coin. The clergyman declared the case to be wickedness, and gave the luckless boy a thrashing; but the fact of his sleep-talking is presumptive evidence of an abnormal state of brain, and though I could not learn anything further of the boy, I am convinced that to treat the subject of a diseased or damaged brain with a stick, as though he were a dog, is in the highest degree mischievous.

I recommend you to give this subject your attention, and carefully consider the difference between criminality and moral insanity, as it is a matter of jurisprudence upon which you may at any time be called upon to offer an opinion.

The ordinary criminal can and does weigh the consequence of his act and the risk of detection, against the benefit, which, in his estimation, will accrue to him through success, and he chooses the risk by a deliberate effort of reason, and he does the act by a deliberate effort of volition, and he deserves and no doubt feels the justice of the punishment he receives, and perhaps, in many instances he might derive benefit from a greater measure. But the unfortunate victim of perverted moral sense or of moral imbecility, has a claim on our pity; he probably, neither feels the punishment inflicted upon him, nor appreciates the degradation of it; and if the punishment does not tend to extinguish all elements of good in this miserable being, it certainly does not benefit him in the least degree.

In concluding this course of lectures I would urge upon you the necessity of studying insanity in conjunction with general disease, rather than alone. Mental disease does not occur without bodily infirmity; enlarge therefore, as much as you can, your basis of mental study. Study the phases of mind in conjunction with all forms of bodily disease, and note the variations in mental state which your patients exhibit under different pathologies, and you will I am sure learn much.

The study of mind has, too long, been trammelled with metaphysical considerations to allow of clear expositions of its various phenomena, but the shackels are now unloosed, and we may return from metaphysical dreams to a solid and material basis. Viewing mind in its relation to matter, and the manifestation of mind in its relation to organization, we may proceed to construct our science according to the synthetical method; nevertheless, bearing in our own consciousness throughout, that the end and aim of our study is the amelioration of suffering and the lessening of that burden which each one has to carry as his allotment of human ills.

APPENDIX.

No. 1. *Form of Medical Certificate.*

MEDICAL CERTIFICATE.—Sched. (A) No. 2

Sects. 4, 5, 8, 10, 11, 12, 13.

I, the undersigned,

being a (*a*)

and being in actual practice as a (*b*)

hereby certify, that I, on the　　　day of　　　at (*c*)

in the County of

separately from any other Medical Practitioner, personally examined　　　　　　　　　　　of (*d*)

and that

the said　　　　　　　　　　　is a (*e*)

and a proper Person to be taken charge of and detained under Care and Treatment, and that I have formed this opinion upon the following grounds, viz :—

1. Facts indicating Insanity observed by myself (*f*)

2. Other facts (if any) indicating Insanity communicated to me by others (*g*)

Signed,　Name　_____

Place of Abode　_____

Dated this　　day of　　One Thousand Eight Hundred and Seventy-

(*a*) Set forth the qualification entitling the person certifying to practise as a physician, surgeon, or apothecary, *ex. gra.* :— Fellow of the Royal College of Physicians in London, Licentiate of the Apothecaries' Company, or as the case may be.

(*b*) Physician, surgeon, or apothecary, as the case may be.

(*c*) Here insert the street and number of the house (if any) or other like particulars.

(*d*) Insert residence and profession, or occupation, (if any), of the patient.

(*e*) Lunatic or an idiot, or a person of unsound mind.

(*f*) Here state the facts.

(*g*) Here state the information, and from whom.

APPENDIX.

No. 1a.

NOTICE OF ADMISSION.

To be forwarded to the Commissioners in Lunacy within one clear day from the Patient's reception.

(a) House or hospital.

I hereby give you Notice, That   was admitted into this *(a)*   as a Private Patient, on the   Day of   187   , and I hereby transmit a Copy of the Order and Medical Certificates on which he was received *(b)*

(b) If a private patient be received upon one certificate only, the special circumstances which have prevented the patient from being examined by two medical practitioners to be here stated, as in the statement accompanying the order for admission.

(c) Superintendent or proprietor of——

Signed _____

(c) _____

Dated this   Day of   One Thousand Eight Hundred and

To the Commissioners in Lunacy.

16 & 17 Vic. c. 96, sched. C., s. 24 (25 & 26 Vic. c. 111.)
Private Patient.

APPENDIX. 439

No. 2. Invalid Certificate, University and College of Surgeons omitted.

MEDICAL CERTIFICATE.—Sched. (A) No. 2.
Sects. 4, 5, 8, 10, 11, 12, 13.

(a) Set forth the qualification entitling the person certifying to practise as a physician, surgeon, or apothecary, ex. gra. :— Fellow of the Royal College of Physicians in London, Licentiate of the Apothecaries' Company, or as the case may be.
(b) Physician, surgeon or apothecary, as the case may be.
(c) Here insert the street and number of the house (if any), or other like particulars.
(d) Insert residence and profession, or occupation (if any), of the patient.
(e) Lunatic or an idiot, or a person of unsound mind.
(f) Here state the facts.
(g) Here state the information and from whom.

I, the undersigned, being a (a) *Doctor of Medicine and a Surgeon* and being in actual practice as a (b) *Surgeon*, hereby certify, that I, on the 1st day of *June*, 1872, at (c) 2 *King Street, Westminster*, in the County of *Middlesex*, separately from any other Medical Practitioner, personally examined *James Ashford* of (d) 2 *King Street, Westminster, Musician*, and that the said *James Ashford* is a (e) *person of unsound mind* and a proper Person to be taken charge of and detained under Care and Treatment, and that I have formed this opinion upon the following grounds viz.—

1. Facts indicating Insanity observed by myself (f)
   He declared to me that he heard the voice of God, ordering him to kill his children.

2. Other facts (if any) indicating Insanity communicated to me by others (g)

Signed,   *Name*, X. Z., *M.D.*
   Place of Abode,   *Westminster.*

Dated this 1st day of *June*, One Thousand Eight Hundred and *Seventy-two*.

No. 3. Invalid Certificate, particular College omitted.

MEDICAL CERTIFICATE. — Sched. (A) No. 2.
Sects. 4, 5, 8, 10, 11, 12, 13.

I, the undersigned, being a (a) *Fellow of the Royal College of Surgeons,* and being in actual practice as a (b) *Surgeon,* hereby certify, that I, on the 1st day of *June,* 1872, at (c) 1 *George Street, St. James's,* in the County of *Middlesex,* separately from any other Medical Practitioner, personally examined *William Fox,* of (d) 1 *George Street, St. James's, Banker,* and that the said *William Fox* is a (e) *Lunatic* and a proper Person to be taken charge of and detained under Care and Treatment, and that I have formed this opinion upon the following grounds, viz.—

1. Facts indicating Insanity observed by myself (f)

    He is labouring under the delusion that he has lost all his money in his business, and that he is bound to cut his throat in consequence.

2. Other facts (if any) indicating Insanity communicated to me by others (g)

    John Fox, his brother, informed me, that instead of having lost money, the patient's business was never in a more prosperous condition.

Signed,   Name, *W. B., F.R.C.S.,*
               Place of Abode, *St. James's.*

Dated this 1st day of *June,* One Thousand Eight Hundred and *Seventy-two.*

(a) *Set forth the qualification entitling the person certifying to practise as a physician, surgeon, or apothecary, ex. gra. :—* Fellow of the Royal College of Physicians in London, Licentiate of the Apothecaries' Company, *or as the case may be.*

(b) *Physician, surgeon, or apothecary, as the case may be.*

(c) *Here insert the street and number of the house (if any), or other like particulars.*

(d) *Insert residence and profession, or occupation (if any), of the patient.*

(e) *Lunatic, or an idiot, or a person of unsound mind.*

(f) *Here state the facts.*

(g) *Here state the information and from whom.*

No. 4. Invalid Certificate, number of house omitted.

MEDICAL CERTIFICATE.—Sched. (A) No. 2.
Sects. 4, 5, 8, 10, 11, 12, 13.

(a) Set forth the qualification entitling the person certifying to practise as a physician, surgeon, or apothecary, ex. gra.:— Fellow of the Royal College of Physicians in London, Licentiate of the Apothecaries Company, or as the case may be.

(b) Physician, surgeon, or apothecary, as the case may be.

(c) Here insert the street and number of the house (if any), or other like particulars.

(d) Insert residence and profession, or occupation (if any), of the patient.

(e) Lunatic, or an idiot, or a person of unsound mind.

(f) Here state the facts.

(g) Here state the information and from whom.

I, the undersigned, being a (*a*) *Doctor of Medicine of the University of Oxford* and being in actual practice as a (*b*) *Physician*, hereby certify, that I, on the 1*st* day of *June*, 1872, at (*c*) *Cavendish Square*, in the County of *Middlesex*, separately from any other Medical Practitioner, personally examined *James Miller* of *(d) Cavendish Square, no occupation,* and that the said *James Miller* is a (*e*) *person of unsound mind*, and a proper Person to be taken charge of and detained under Care and Treatment, and that I have formed this opinion on the following grounds, viz :—

1. Facts indicating Insanity observed by myself *(f)*
  He is under the delusion that he is the subject of a conspiracy, and that he is perpetually watched and followed by detectives.

2. Other facts (if any) indicating Insanity communicated to me by others (*g*)
  Jessie Miller, his sister, told me that he becomes suddenly violent, and declares that people in the street are keeping watch on him.

Signed, *Name E. G., M.D.* (Oxon.),
Place of Abode, Harley Street, W.

Dated this 1*st* day of *June,* One Thousand Eight Hundred and *Seventy-two.*

No. 5. *Valid Certificate.*

MEDICAL CERTIFICATE. — Sched. (A) No. 2.
Sects. 4, 5, 8, 10, 11, 12, 13.

I, the undersigned, being a (*a*) *Member of the Royal College of Physicians in London*, and being in actual practice as a (*b*) *Physician* hereby certify, that I, on the *sixth* day of *May* 1872, at (*c*) 00 *Seymour Street, Portman Square*, in the County of *Middlesex*, separately from any other Medical Practitioner, personally examined *Dorothea Elden* of (*d*) 00 *Seymour Street, Portman Square, Gentlewoman*, and that the said *Dorothea Elden* is a (*e*) *person of unsound mind*, and a proper Person to be taken charge of and detained under Care and Treatment, and that I have formed this opinion upon the following grounds, viz.—

1. Facts indicating Insanity observed by myself (*f*)
She is the subject of delusional Insanity, and stated to me that she constantly heard the voice of the Devil, who influenced the people in the house to annoy her, and that she was prevented from sleeping by rows caused in the streets by the Devil.

2. Other facts (if any) indicating Insanity communicated to me by others (*g*).
Her sister-in-law, Mrs. Catherine Haynes, informed me that the patient burst out laughing in church during service on Sunday last, and that her conversation, formerly always modest, has now become at times improper, and that she has become disaffected towards her mother.

Signed, *Name W. F. C.*,
Place of Abode, Great Cumberland Place.

Dated this *sixth* day of *May*, One Thousand Eight Hundred and *Seventy-two*.

(*a*) Set forth the qualification entitling the person certifying to practise as a physician, surgeon, or apothecary, ex. gra.:— Fellow of the Royal College of Physicians in London, Licentiate of the Apothecaries Company, *or as the case may be.*
(*b*) Physician surgeon, *or* apothecary, *as the case may be.*
(*c*) Here insert the street and number of the house (if any), or other like particulars.
(*d*) Insert residence and profession, or occupation (if any), of the patient.
(*e*) Lunatic, or an idiot, or a person of unsound mind.
(*f*) Here state the facts.

(*g*) Here state the information and from whom.

APPENDIX. 443

No. 6. *Certificate with Erasures Initialed.*
MEDICAL CERTIFICATE.— Sched. (A) No. 2.
Sects. 4, 5, 8, 10. 11, 12, 13.

(a) Set forth the qualification entitling the person certifying to practise as a physician, surgeon or apothecary, ex. gra. :— Fellow of the Royal College of Physicians in London, Licentiate of the Apothecaries' Company, or as the case may be.
(b) Physician, surgeon, or apothecary, as the case may be.
(c) Here insert the street and number of the house (if any), or other like particulars.
(d) Insert residence and profession, or occupation (if any), of the patient.
(e) Lunatic, or an idiot, or a person of unsound mind.
(f) Here state the facts.

(g) Here state the information and from whom.

I, the undersigned, being a (a) ▬▬▬▬ *Mem-* *E.H.* *ber of the Royal College of Physicians in London,* and being in actual practice as a (b) *Physician*, hereby certify, that I, on the *sixth* day of *May* 1872, at (c) 5O *George Street, Hanover E.H. Square*, in the County of *Middlesex* separately from any other Medical Practitioner, personally examined *Mary Anne Farthing* of (d) *Ellisland, Carlisle, Cumberland, farmer's daughter*, and that the said *Mary Anne Farthing* is a (e) *person of unsound mind* and a proper Person to be taken charge of and detained under Care and Treatment, and that I have formed this opinion on the following grounds, viz.—

1. Facts indicating Insanity observed by myself *(f)*

2. Other facts (if any) indicating Insanity communicated to me by others (*g*)

Signed, Name, E. H.,
Place of Abode, 50 *George Street, Hanover Square.*

Dated this *6th* day of *May*, One Thousand Eight Hundred and *Seventy-two*.

APPENDIX.

No. 7. Invalid Order and Statement.

ORDER FOR THE RECEPTION OF A PRIVATE PATIENT.
Sched. (A) No. 1. Sects. 4, 8.

I, the undersigned, hereby request you to receive *Miss Mary Ann C——*, whom I last saw at 16 *Westmoreland Place, Hyde Park, County of Middlesex*, on the *third* day of *January, last*, 1872, (*a*) a (*b*) as a Patient into your House.

(*a*) Within one month previous to the date of the order.
(*b*) Lunatic, *or* an idiot, *or* a person of unsound mind.

Subjoined is a Statement respecting the said

Signed, Name, *George Arthur C——*
Occupation (*if any*), *Clerk in Holy Orders,*
Place of Abode —— *Vicarage,* ——*Kent.*
Degree of Relationship (*if any*), or other circumstances of connexion with the Patient. } ·1*st Cousin.*

Dated this *Nineteenth* day of *January*, One Thousand Eight Hundred and *Seventy-two*.

To _____

(*c*) Proprietor *or* superintendent of——
(*d*) Describing the house, *or* hospital by situation and name (if any.)

(*c*) _____ (*d*) _____

STATEMENT.

If any Particulars in this Statement be not known, the fact to be so stated.

Name of Patient, with Christian name at length	*Mary Ann C.*
Sex and Age	*About* 60 *or* 65.
Married, Single, or Widowed . .	*Single.*
Condition of Life, and previous Occupation (if any)	*No occupation.*
Religious Persuasion, as far as known	*Church of England.*
Previous Place of Abode . . .	*Was removed from a boarding house in London to one in Berkshire, as supposed; has lived much in boarding houses or lodgings.*

APPENDIX. 445

Whether First Attack . . .
Age (if known) on first Attack . .
When and where previously under Care
   and Treatment . . . .
Duration of existing Attack . . .
Supposed Cause . . . . .
Whether subject to Epilepsy . . .
Whether Suicidal . . . . .   *Not to my knowledge ; but deemed proper by me that she should not be left alone.*

Whether Dangerous to others . . .   *Do.*
Whether found Lunatic by Inquisition
   and Date of Commission or Order for
   Inquisition . . . . .
Special Circumstances (if any) preventing the Patient being examined, before Admission separately by Two Medical Practitioners . . . .
Name and Address of Relative to whom   *Rev. George Arthur C. ——Vicarage*
Notice of Death to be sent . . .   *——Kent.*

(e) Where the person signing the statement is not the person who signs the order, the following particulars concerning the person signing the statement are to be added.

Signed, Name, (e) _____
Occupation (if any) _____
Place of Abode, _____
Degree of Relationship (if any) ⎫
or other circumstance of con- ⎬ _____
nexion with the Patient ⎭

———

No. 8. Ordinary Form.

ORDER FOR THE RECEPTION OF A PRIVATE PATIENT.

Sched. (A) No. 1, Sects. 4, 8.

———

I, the undersigned, hereby request you to receive

(a) Within one month previous to the date of the order.
(b) Lunatic, or an idiot, or a person of unsound mind.

whom I last saw at
on the   day of (a)
a (b)   , as a Patient into your House.

Subjoined is a Statement respecting the said

    Signed, Name _____
    Occupation (if any) _____
    Place of Abode _____

Degree of Relationship (if any),
or other circumstances of con- } _____
nexion with the Patient

(c) Proprietor, or superintendent of——
(d) Describing the house or hospital by situation and name, if any.

Dated this   Day of
One Thousand Eight Hundred and Seventy-
To _____

    (c) _____ (d) _____

STATEMENT.

If any Particulars in this Statement be not known, the Fact to be so stated.

Name of Patient, with Christian Name at length .
Sex and Age .
Married, Single, or Widowed
Condition of Life, and previous Occupation (if any)
Religious Persuasion, as far as known .
Previous Place of Abode
Whether First Attack .
Age (if known) on First Attack
When and where previously under Care and Treatment
Duration of existing Attack .
Supposed Cause
Whether subject to Epilepsy .
Whether Suicidal .
Whether Dangerous to others
Whether found Lunatic by Inquisition, and Date of Commission or Order for Inquisition .
Special Circumstances (if any) preventing the Patient being examined, before Admission, separately, by Two Medical Practitioners
Name and Address of Relative to whom Notice of Death to be sent

(c) Where the person signing the statement is not the person who signs the order, the following particulars concerning the person signing the statement are to be added.

    Signed, Name (e) _____
    Occupation (if any) _____
    Place of Abode _____

Degree of Relationship (if any),
or other circumstances of con- } _____
nexion with the Patient

APPENDIX. 447

No. 9. *Ordinary Form.*
PRIVATE PATIENT.—(25 & 26 Vic. c. 111, § 28.)

STATEMENT.

(*a*) After two clear days, and before the expiration of seven clear days, from the admission of the patient.

I have this Day (*a*) seen and examined   a Private Patient, received on the   day of   18   pursuant to an Order dated the   day of   18   and hereby Certify, that with respect to Mental State   he

and that with respect to Bodily Health and Condition he

Signed,

(*b*) Medical Proprietor, *or* Superintendent, *or* Attendant of

(*b*)

Dated this   Day of
One Thousand Eight Hundred and Sixty.

No. 10. *Ordinary Form.*
FORM OF MEDICAL VISITATION BOOK OR MEDICAL JOURNAL.

DATE:	Mental State and Progress.	Bodily Health and Condition	Restraint or Seclusion since last entry. When and how long? By what means, and for what reasons?	Visits of Friends	State of House, Bed, and Bedding, &c.

No. 11. *Ordinary Form.*
FORM OF NOTICE OF DEATH.

I hereby give you notice, That a Private Patient, received into this house on the   day of   18   , died therein on the   day of   187   , and I further Certify, that   was present at the Death of the said   and that the apparent Cause of Death of the said   (*a*) was

Signed, _____
(*b*)_____

Dated this   day of   One Thousand Eight Hundred and Seventy

To the Commissioners in Lunacy.

(*a*) Ascertained by post-mortem examination, *if so.*
(*b*) Medical Proprietor of ——— house, or Medical Attendant.

No. 12. *Ordinary Form.*
FORM OF NOTICE OF DISCHARGE.

I hereby give you Notice, That a Private Patient, received into this house on the day of   18   was discharged therefrom (*a*) by the Authority of   on the   day of   187

Signed,
(*b*)_____

Dated this   day of   One Thousand Eight Hundred and Seventy

To the Commissioners in Lunacy.

(*a*) Recovered, *or* relieved, *or* not improved.
(*b*) Proprietor of——— house.

No. 13.
TRANSFER OF PRIVATE PATIENT.
(16 & 17 Vic., Cap. 96, § 20.)

CONSENT.

We, the Undersigned, Commissioners in Lunacy, hereby consent to the removal, on or before the   day of
187   , of   a Private Patient,
in   House,
to   House,

Given under our hands this   day of
in the year of Our Lord One Thousand Eight Hundred and Seventy

────────────────────────── ⎫
                           ⎬ *Commissioners in Lunacy.*
────────────────────────── ⎭

ORDER.

I,*   the undersigned having Authority to discharge   a Private Patient in   House,

hereby order and direct that the said
be removed therefrom to   House,

Given under my hand this*   day of
in the year of Our Lord One Thousand Eight Hundred and Seventy

Signed,
Place of Abode

---

* NOTE. This order must be signed and dated *subsequently* to the consent of the Commissioners; and it must be signed by

Generally
1. The person *who signed the order* for the Patient's admission:
2. If such person be incapable (by reason of insanity, or absence from England, or otherwise), or if he be dead, then by the *Husband or Wife* of the patient:
3. If there be no Husband or Wife, then by the Patient's *Father:*
4. If there be no Father, then by the Patient's *Mother:*
5. If there be no Father or Mother, then by any *one* of the Patient's *nearest of kin:* Or by the person *who made the last payment* on the Patient's account.

In cases of Chancery Patients—The Committee of the Person.

APPENDIX.

No. 14.
ORDER FOR THE RECEPTION OF A PAUPER PATIENT.

Sched. (B.) No. 1, Sect. 7.

(a) I, *A. B.*, a justice of the peace for the county, city, or borough of——; *or in the case of a clergyman and relieving officer or overseer, We C. D. and E. F.*

(b) Physician, or surgeon, or apothecary, *as the case may be.*

(c) Lunatic, or an idiot, or a person of unsound mind.

(d) Justice of the peace for the county, city, or borough of——; *or an or the officiating clergyman of the parish of——.*

(e) The relieving officer of the union or parish of——, *or an overseer of the parish of——*

(a) the undersigned, having called to   Assistance a (b)
      and having personally examined
a Pauper, and being satisfied that the said
                is a (c)
and a proper Person to be taken charge of and detained under Care and Treatment, hereby direct you to receive the said
               as a Patient into your House.
Subjoined is a Statement respecting the said
  Signed, Name (d) _____

                                        *Name (e)*_____

Dated the   Day of
One Thousand Eight Hundred and
To

STATEMENT.
If any Particulars in this Statement be not known, to be so stated.

Name of Patient, and Christian Name at length . .
Sex and Age . . . .
Married, Single, or Widowed . . . .
Condition of Life, and previous Occupation (if any) .
Religious Persuasion, as far as known . . .
Previous Place of Abode . . . . .
Whether First Attack . . . . .
Age (if known) on First Attack . . . .
When and where previously under Care and Treatment
Duration of existing Attack . . . . .
Supposed Cause . . . . . .
Whether subject to Epilepsy . . . . .
Whether Suicidal . . . . . .
Whether Dangerous to others . . . .
Parish or Union to which the Lunatic is chargeable .
Name and Christian Name, and Place of Abode of
  nearest known Relative of the Patient, and Degree
  of Relationship, if known . . . . .

        I certify that, to the best of my knowledge, the above Particulars are correctly stated.
         Signed, Name _____
                 (f)_____

(f) Relieving officer or overseer.

GG 2

No. 15.
MEDICAL CERTIFICATE.
Sched. (B) No. 2, Sects. 7, 10, 11, 12, 13.

I, the undersigned, being   and being in
actual practice as a (*a*)   hereby
certify, that I, on the   day of
at (*b*)   in the
County of
personally examined (*c*)
of   Street,
and that the said
is a (*d*)   and a proper Person to be
taken charge of and detained under Care and Treatment, and that I have formed this opinion upon the following grounds, viz.—

(*a*) Physician, surgeon, *or* apothecary, *as the case may be.*
(*b*) Here insert the street and number of the house (if any), or other like particulars.
(*c*) Insert residence and profession, or occupation (if any) of the patient.
(*d*) Lunatic, *or* an idiot, *or* a person of unsound mind.

1. Facts indicating Insanity observed by myself (*e*)

(*e*) Here state the facts.

2. Other facts (if any) indicating Insanity communicated to me by others (*f*)

(*f*) Here state the information, and from whom.

Signed, *Name*,_____
Place of Abode,_____
Dated this   Day of
One Thousand Eight Hundred and

No. 16.

NOTICE OF ADMISSION.
Sched. (C) Sect. 24.

I hereby give you Notice, That   was admitted into this House as a Pauper Patient, on the   Day of   18   ,
and I hereby transmit a Copy of the Order and Medical Certificate on which   he was received.

Subjoined is a Statement with respect to the Mental and Bodily Condition of the above-named Patient.

Signed, _____

Medical Superintendent,
of

Dated this   Day of
One Thousand Eight Hundred and

To the Commissioners in Lunacy.

STATEMENT.

(*b*) Some day not less than two clear days after the admission of the patient.

I have this Day (*b*) seen and examined   the Patient mentioned in the above Notice, and hereby Certify, that with respect to Mental State   he
and that with respect to Bodily Health and Condition   he

Signed, _____

Medical Proprietor, or Superintendent
of

Dated this   Day of
One Thousand Eight Hundred and

INDEX.

Abnormal conditions of Sympathetic, 417.
Accidents, 138, 283.
Acute diseases, 146.
Age in progressive paralysis, 269.
Agraphia, 249.
Ague, 148.
Alcohol in mania, 167.
Alienist's view of General Paralysis, 243.
Alimentary tract delusion in association with, 205.
Ammonium bromide of, 175.
Amyl nitrite of, 175.
Anatomy, pathological of the Insane, 385.
Antimony, 166.
Aphasia, 375.
Artificial mode of living, 385.
Asomnia, 82, 84, 158.
Asylum treatment, necessity for, 388.
Ataxy progressive, locomotor, 264.
cord in, 275.
Attention, disorder of faculty of, 77.
Attendants, 163.
Atrophy, of Brain and Cord, 277, 278.
progressive muscular, 277.

Aurum Potabile, 218.
Axis, cylinders, 276.

Bain, Alexander, 418.
Ballarger, 235.
Baron Martin. (quoted) 41.
Bastian, Dr. Charlton, (quoted) 102.
Bateman, Dr. Frederick, (quoted) 375.
Baths,
in mania, 158, 170.
in melancholia, 229.
in paralysis, 287.
Bayle, 214, 234.
Bean, Calabar, 286.
Beale's Archives, 154.
Belladonna, 175.
Blandford, Dr., (quoted) 73, 102, 213, 230, 406.
Bleeding in mania, 166.
Blisters, 287.
Blocked Vessels, 103.
Blood Supply, diminution of, 213, 375.
Blows, 283, 362.
on the Head, 98.
Blushing, 418.
Boarding out system, 393.
Bodily state of imbecility, 335.
Boils, 16.
Boismont, Brierre de, 235.
Bones, fragility of, 287.
Bony tubercles, 155.
Bonnet, Dr. Henri, 271, 280.
Boyd, Dr., (quoted), 146, 235, 271, 273, 287.

INDEX.

Brain, Œdema of, 279.
  in melancholia, 209.
  microscopical, pathology of, 271.
  in progressive locomotor ataxy, 275.
  progressive paralysis, 271.
Bricheteau, M., (quoted) 372.
Browne, Dr. Crichton, (mentioned) 87, 271.
Brown-Sequard, (quoted) 126-7.
Bucknill, Dr., (quoted) 87.
Burrows, Dr., (quoted) 300, 371, 372.
Burton, (quoted) 177, 218.
Byles, Mr. Justice, (quoted). 41.
Byron, (quoted) 195.

Cabanis, theory of, 4.
Calabar Bean, 286.
Cannabis Indica, 166, 232.
Canine Madness, 206.
Calmeil, 214, 235.
Capillary Apoplexy, 102.
Capacity to make a will, 252.
Carbuncle, 16.
Cardiac disease, 140, 231.
Catalepsy, 131.
  in Men, 133.
Caudate Cells, 265, 277.
Cause of death in paralysis, 285.
Causes of
  mania, 92, 105.
  melancholia, 212.
  progressive paralysis. 283.
  acute dementia, 300.
Cerebral Congestion, 103.
Certificates of Lunacy, 44.
  invalid, 46, 47.

Cessation of Fluxes, 90.
Change of Scene, 232, 386.
Changes in Sympathetic, 280.
Chancery Enquiry, 42.
Cheerful Rooms, 222.
Child, Dr. Gilbert, 345.
Children,
  morbid state of mind in, 373.
Chloral, 167.
Chloroform, 378.
Chronic delusions, 107, 112.
  dementia, 308.
  mania, 105.
  melancholia, 199.
  inflammation of
    brain, 279.
    cord, 278.
    membranes, 279.
Circulation,
  influence of, 371.
Clarke, Dr. Lockhart, (quoted), 152, 153, 154, 210, 235, 246, 271, 272, 273.
Classes, affected by progressive paralysis, 285.
Classification, 408.
  outline of, 15.
Climacteric period, 141.
Clinical observation, 88.
Clouston, Dr. (quoted), 145, 410.
Coleridge, Mr. Justice, (quoted), 47.
Coleridge, Samuel Taylor, 396.
Combe, Dr. A. (quoted), 372.
Condition of Convolutions in Idiots, 332.
  lips in Imbeciles and Idiots, 328.
  mouth and teeth in Idiots, 340.

INDEX. 457

soft palate, in Idiots 337.
wasting of brain, 315.
Conolly, Dr., (mentioned), 81, 86, 180, 187.
Conclusion, 436.
Constitutional disease, 96, 231.
Copies of certificates, 59.
  entries, 65.
Cord, Spinal,
  axis cylinder, 276.
  cells of, in progressive muscular atrophy, 277.
  chronic inflammation of, 278.
  in progressive locomotor ataxy, 275.
  progressive paralysis, generally, 274.
Criminality, and moral insanity, 435.

Dangerous Lunatics, 80.
Darkness, use of in treatment, 169.
Darwin, (quoted), 87.
Death, notice of, 65.
Definition of dementia, 289.
  mania, 7.
  melancholia, 179.
Delirium, 407.
Delirium Tremens, 19.
Delusion, 75, 205, 400.
  as evidence of insanity, 401.
  of greatness, 247.
Dementia, 289.
  acute, 292, 299.
  chronic, 308.
  infantile, 322.
  paralytic, 244.
  primary, 316.
  progressive, 316.
  senile, 320.
  treatment of, 308.
  in progressive paralysis, 238, 249.

Demonomaniacs, 135.
Departure from ordinary conduct, 405.
Development of Skull in Idiots, 332.
Destructive habits, 172.
Diagnosis, 18.
  of delirium tremens, 23.
  of dissembling, 26, 116.
  of epilepsy, 27.
  of eccentricity, 19, 28.
  of fever, 19.
  history, value of, in, 25.
  of hysteria, 19.
  of malingering, 19, 25.
  of melancholia, 216.
  overt acts, value of, in, 19.
  pulse, value of, in, 20.
  temperature, value of, in, 20.
  tongue, value of, in, 20.
Diamond, Dr., 86.
Dickinson, Dr., (quoted), 152.
Digitalis, in mania, 165.
Diminution of blood supply, 213, 375.
Disaffection, 163.
Discharge, order of, 66.
  who has power to, 66.
Disease, acute, 146.
  mental, 1.
  of viscera in melancholia, 214.
Dog madness, 206.
Down, Dr. Langdon, (quoted), 323, 329, 338, 339, 340, 349, 350, 351, 354.
Dryden, (quoted), 187.
Dura Mater,
  inflammation of brain, 101.
            of cord, 278.
Duvay, Dr., 345.

Eccentricity, 19, 28.
Education, 359.
Effect of vasi motor nerves, 419.
Emotions, 413.
  expression of, 87.
  influence of, 422.
  insanity of, 416.
Ependyma, dropsy of, 324.
Epilepsy, 27, 117.
  nocturnal seizures, 133.
  in progressive paralysis, 251, 266.
Epileptics, violence of, 135.
Epileptiform Seizures, in progressive paralysis, 281.
Erasures, 48.
Esquirol, (quoted), 178, 290, 409.
Escape, notice of, 65.
Ethnical classification of Idiots, 349.
Etiology of dementia, 323.
  idiocy, 341.
  insanity, 357.
  mania, 92, 105.
  melancholia, 212.
  progressive paralysis, 283.
Evidences of insanity, 33, 394.
Evolution of teeth, 329.
Exaltation in progressive paralysis, 254.
Examination of Patients on admission, 67.
Excitation, 79, 84.
Exercise, 382.
Exhaustion, 93.
Expression in insanity, 86.
  of emotions, 87.
Expressions, 246.

Failure of nerve power, 247.

Falret, Jules, (quoted), 118, 121, 124, 127, 235.
Feeding, 223.
  nose, 225.
  spoon, 223, 226.
  stomach pump, 225.
Firmness, necessity for, 220.
Flow, lochial, 88, 90.
Fluid, metamorphosis of cells, 277.
Flushing of face, 128.
Fluxes, reappearance of, 90.
Folie, paralytic, 234.
  circulaire, 253, 258.
Food, necessity for, in melancholia, 223.
Forgetfulness, 246.
Formation changes, 151.
Functional disturbance, 417.

Gaining access to patients, 44.
Ganglia,
  of posterior roots, 278.
  sympathetic changes in, 281.
General Note on Treatment, 386.
  on pathological anatomy, 385.
General Paralysis of the Insane, 234.
  various views of, 234.
  objections to the term, 234.
Gout, 147.
Granulations of Ependyma, 154.
Greenwood, case of, 47.
Griesinger (quoted), 114, 400.
Ground floor rooms, value of, 162.
Grounds of Study, 6.
Gull, Sir William, (quoted), 197, 367.

Hæmatozin in Brain,
in melancholia, 210.
in progressive paralysis, 210.
Hair in mania, 87.
Hale, Chief Justice, (quoted), 38.
Hallucination, 35, 397, 401.
waking, 358.
Harsten, Dr. F. A., (mentioned), 101.
Haughton, Rev. Professor, (quoted), 94.
Heart Disease in melancholia, 212.
Henbane in Mania, 166.
Hereditary Predisposition, 93.
Hill, Dr. Gardner, (quoted), 81.
History, value of, 25.
Hippocrates, 177.
Home Treatment, 157, 160, 389.
Howe, Dr., 342, 345.
Human Soul, 2.
Hutchinson, Mr. Jonathan, 341.
Hydrocyanic Acid, 166.
Hydrophobia, 207.
Hyperæmia, 151.
Hypostatic Pneumonia, 285.
Hysteria, 19, 29, 416.

Idiocy and Imbecility, 327.
Idiocy, defined, 327.
etiology of, 341.
Idiots
convolutions in, 332.
ethnical classification of, 349.
insentient, 330.
mental power of, 331.
size of head, 328.

Idiots and Imbeciles
condition of lips of, 328.
tongue, 329.
palate, 329.
evolution of teeth in, 329.
Idiopathic epilepsy, 368.
Illusion, 394.
Imbeciles,
mental faculties of, 334.
size of heads, 328.
Imbecility defined, 327.
Impaired nutrition, 151.
Imperfect recovery, 105.
Impulse on waking, 190.
of violence, 403.
Incongruities of dress, 405.
Initials, 48.
Infantile dementia, 322.
Inflammation of dura mater, 101, 150, 279.
of brain, 279.
of cord, 278.
Insanity,
etiology, 356.
evidence of, 33, 394.
metastatic, 144.
popular notions of, 29.
progressive paralysis and, 247.
puerperal, 143.
Insolatio, 139.
Introduction, 1.
Invalid certificates, 46, 47.
Iodide of Potassium, 174, 286.

Jackson, Dr. J. Hughlings, (quoted), 370, 375, 377.
Jamieson, Dr., 394.
Jepson, Dr., (mentioned), 94, 377.
Johnson, Dr., 28.
Joire, M., (quoted), 271.

Kolk, Schrœder Van der, (quoted), 11, 101, 127, 145, 150, 154, 156, 166, 175.
Kussmaul and Tenner, (quoted), 126, 127.

Lacrymose melancholia, 202.
Language, necessity for, 339.
Laycock, Professor, 415.
Legal Relations of Lunatics, responsibilities, 37.
definitions, 38.
dictum, 38.
Licenses, 68.
Local effects of diminution of blood supply, 375.
Lochial flow, 88, 90.
Locomotor ataxy, progressive, 264.
Longevity, in relation to brain work, 18.
Loquacity, 109.
Lord Chancellor, 42.
„ Chancellor's Visitors, 43.
„ Lyndhurst (quoted), 39.
Loss of Words, 374, 375.
music, 249.
Lunacy Commission, 43.
Lunatics
responsibility of, 36.
wandering, 62.
Lyndhurst, Lord, (quoted), 39.
Lypemania, 178.

McNaughten, case of, 38.
Macnish quoted, 133.
Madness,
medical opinions of, 29.

Major, Dr. Herbert, 272.
Male and female attendant, 163.
Malingering, 25.
Mal-nutrition, 137, 151.
Mania, acute, 70.
classes of, 73
chronic, 105.
definition of, 70, 73.
etiology of,
acute, 92.
chronic, 105.
hair in, 87.
pathology of, 99, 149.
recurrent, 138.
varieties of, 117, 137.
Maniacs,
dangerous, 80.
Manie des grandeurs, 234, 247.
Marriages of Consanguinity, 344.
of tainted, 366.
of epileptics, 367.
Martin, Baron, (quoted,) 41.
Martin, Theodore, 382.
Masturbation, 141, 175.
Maudsley, Dr., (quoted,) 153, 213, 347, 343, 365, 410, 426.
Maxwell, Dr., 353, 354.
Medical,
attendant, 61.
journal, 65.
visitation book, 64.
Medicine, Psychological, 2.
Melancholia, 177.
acute, 179.
baths in, 229.
Cannabis Indica in, 232.
change of scene, 232.
chronic, 199.
cold baths in, 229.
comparison with mania, 207.
condition of, 210.

INDEX. 461

Melancholia.
  constitutional and local disease, 231.
  definition of, 179.
  derivation of the term. 177.
  diagnosis of, 216.
  disease of heart in, 212.
    other viscera, 214.
  etiology of, 212.
  hallucination in, 191.
  heart disease, 212.
  homicidal tendences, 195.
  microscopical conditions in, 209.
  narcotics in, 230.
  pathology of, 203, 209.
  physical condition of, 203.
  purgatives, 228.
  restraint of feeling in, 200.
  sedatives, 230.
  stage of recovery, 233.
  state of mal-nutrition, 204.
  stimulants in, 231.
  syphilis in connection with, 215.
Mellor, Mr. Justice, (quoted,) 41.
Memory, 76.
Meningitis, Tubercular, 324.
Menorrhagia in melancholia 231.
Menstruation, 143.
Mental disease, 1.
Mental powers of Idiots, 331.
  faculties of Imbeciles, 334.
Metamorphosis fluid of cord, 277.
Metastatic insanity, 144.
Microscopical anatomy, conditions, 10.
  pathology of brain, 271.
    cord, 274.
Mill, J. S., (quoted,) 432.

Mind, morbid states of, 394.
Mistakes of sense, 414.
  classified, 415.
Mitchell, Dr. Arthur, 344.
Moral causes of insanity, 380.
  sense and insanity, 423.
  Imbeciles, 334, 427.
Morbid processes in insanity, 16.
  states of mind, 394.
  in children, 373.
Morbus cordis, 140, 213, 300.
Morel, (quoted,) 347, 348, 365, 379.
Muscular atrophy, progressive, 277.
  paralysis in paralytic dementia, 265.
Music, loss of, 249.
Moore, Mr. C. Hewitt, 225,
Morphia, 357.
  bimeconate, 230.
Moxey, Dr. A., 225.
Moxon, Dr., 266, 293, 375.

Narcotics in melancholia, 230.
Natural classification, 412, 413.
Necessity for asylum treatment, 388.
  food, 223.
  language, 339.
  sleep, 164.
Nerve roots, 278.
Nerves, vasi motor effect of, 419.
Nervous system,
  Hospital for diseases of the, 133.
  increase of
  diseases of, 368.
  syphilitic diseases of, 369.

Nitrite of amyl, 175.
Nocturnal epileptic seizures 133.
Non-restraint, 81.

Object of Study, 2.
Objections to Alcohol, 169.
Observations clinical, in insanity, 88.
Œdema of Brain, 279.
Opalescent Arachnoid, 101.
Opium, 164.
Order and Statement, 57.
Outline of
  classification, 15.
  etiology, 12.
  general pathological anatomy, 8, 16,
  pathology, 11.
  prognosis, 34.

Pachymeningitis of brain, 150.
  of cord, 279.
Paresis, 234.
Pathology,
  outline of, 11.
  acute mania, 99.
  chronic mania, 149.
  acute dementia, 299.
  chronic dementia, 314.
  melancholia, 203.
  progressive paralysis, 270.
  puerperal insanity, 90.
  progressive locomotor ataxy, 275.
Paralysis agitans, 265.
Pathological specimens, 150.
Patients gaining access to, 44.
  regulations for reception of, 58.
  in private houses, 60.
  by friends, 61.

Period climacteric, 141.
Persons prohibited from signing certificates, 56.
Petit Mal, 27, 119.
Phthisis, 144.
Philosophy, 6.
Philpots, Dr. E. P., 207.
Pope, (quoted,) 179.
Popular notions of insanity, 29.
  objections to medical views of madness, 29.
  errors, 406.
Post mortem conditions in chronic mania, 150.
Poincarè, Dr., 271, 280.
Potentiality, 13, 360, 362.
Power of understanding, 80.
Puberty, 140.
Puerperal State, 143.
Pulse, 420.
Predisposition, hereditary, 93.
Pre-occupation of mind, 113.
Primary and secondary conditions, 137.
Progressive paralysis, 235.
Propensities, vicious, 108.
Psychiatry, 2.
Psychological Medicine, 2.

Railway travelling, 378.
Rayner, Dr. (mentioned), 185.
Re-appearance of fluxes, 90.
Reception, 58.
  by friends, 61.
  in private houses, 60.
Recovery, imperfect, 105.
Recurrence of symptoms, 318.
Recurrent mania, 138.
Red softening, 386.

INDEX.

Regulations for reception, 58.
Relation of brain to work, 18.
Relative frequency of progressive paralysis in sexes, 269.
Religious impressions in melancholia, 193.
exercises, 383.
Right and wrong, 37, 423.
Removal from exciting causes, 157.
and transfer, 67
Report, annual, 68.
Rest, 158.
Robertson, Dr. Lockhart, 160.
Robertson's method of wet packing, 160.
Robin, M., 271.
Rokitansky, 210, 247, 271.
Rooms, associated, 221.
cheerful, 222.
single, 221.
Roots of nerves, 278.
posterior, 278.
Rutherford, Dr., 153, 273.

Salomon, Dr., 271.
Sankey, Dr., (mentioned), 210, 235, 247, 271.
Scene, change of, 232.
Sclerosis, 274.
Second part of certificate, 51.
Sedatives in melancholia, 230.
Self-consciousness, 130, 402.
conceit, 403.
Senile dementia, 320.
trembling, 320.
Sequard, Brown, (quoted), 126.
Sexes, relative frequency of paralysis in, 269.

Shakespeare, (quoted), 195.
Shock, 418.
Short description of mania, 80.
Size of head in idiots, 328.
Skae's classification, 410, 411.
Skull, development of, 332.
Sleep, necessity of, 164.
talking, 130.
walking, 130.
Slight physical causes, 97.
Sloughs, 16.
Soft palate in idiots, 337.
Softening, red, 386.
white, 386.
yellow, 386.
Solly, Mr., 332.
Stage of aberration, 135.
recovery in melancholia, 233.
Starvation, 370.
State, mental of idiots, 337.
puerperal, 143.
pupils, 246.
Statement, 57.
Stevens, Dr. Henry, 224.
Stimulants in mania, 103.
melancholia, 231.
Stocker, Dr. A., 3.
Stomach pump feeding, 223.
tube, 224.
Suicidal tendencies, 187.
Suicide only committed in the manner dictated, 191.
Supersedeas, 43, 285.
Surname, 52.
Suspicion, 404.
Sutherland, Dr., (quoted), 213.
Sympathetic depression, 418.
ganglia, 280.
Symptoms, alternations of, 253, 258, 318.
Syphilis, 146.
as a cause, 215, 369.

Syphilitic teeth, 341, 369.
  diseases of the nervous system, 369.
System, boarding out, 393.

Tabic and Paralytic forms of progressive paralysis, 263.
Tenner, Kussmaul and, (quoted), 126, 127.
Temperature, 75.
Thecal abscesses, 16.
Theory of Cabanis, 4.
Thickening of membranes, 100.
Treatment, outline of, 35.
  dementia, 308, 325.
  general note on, 386.
  idiocy and imbecility, 352.
  mania, 157.
  melancholia, 217.
  paralysis, 286.
Tremulous lips, 246.
  tongue, 246.
Trousseau (quoted), 122, 124, 127, 130.
Tubercles of bone, 155.
Tubercular Meningitis, 324.
Tuberculosis, 136, 307.
Tuke, Dr. Batty, 143, 153.
  Dr. Daniel, (quoted), 194, 201, 312.
  Dr. Harrington, 225.

Tuke, Dr. William, (mentioned), 80.

Vacillation of symptoms, 258.
Virchow, (mentioned), 332.
Visitors, Lord Chancellor's, 13, 43.

Walking, sleep, 130.
Waking, impulse on, 190.
  hallucination, 358.
Want and privation, 94.
Warm baths, 158.
Wasting changes, 152, 274.
  of brain, 152.
  of cord, 278.
  of nerve roots, 278.
Wedl, 210, 247, 271.
Westphal, C. (quoted), 127, 235, 248, 254, 263, 264, 273.
Wilks, Dr., (quoted), 196, 235, 243, 244, 245, 246, 249, 364.
Williams, Dr. Duckworth, 226.
Winslow, Dr. Forbes, (quoted), 123, 380.

Yellowlees, Dr., 364.

www.ingramcontent.com/pod-product-compliance
Lightning Source LLC
Chambersburg PA
CBHW021423300426
44114CB00010B/617